Solar Energy
Thermal Processes

SOLAR ENERGY THERMAL PROCESSES

by

JOHN A. DUFFIE
Professor of Chemical Engineering

and

WILLIAM A. BECKMAN
Professor of Mechanical Engineering

Solar Energy Laboratory
University of Wisconsin — Madison
Madison, Wisconsin

A WILEY-INTERSCIENCE PUBLICATION

JOHN WILEY & SONS   New York · London · Sydney · Toronto

Library of Congress Cataloging in Publication Data:

Duffie, John A
  Solar energy thermal processes.

  "A Wiley-Interscience publication."
  Includes bibliographies.
  1. Solar energy. 2. Heat engineering. I. Beckman, William A., joint author. II. Title.

TJ810.D8   621.47   74-12390
ISBN 0-471-22371-9

Printed in the United States of America

10

Dedicated to Pat and Sylvia

PREFACE

Renewed interest in solar energy has developed since 1970 as a result of increasing costs of energy from conventional resources and the problems of importing and extracting fuels that are acceptable from environmental standpoints. The engineering design and analysis of solar processes present unique problems, due to the intermittent and diffuse nature of the resource and the capital-intensive (high initial cost) nature of the processes. The purpose of this book is to summarize the state of knowledge relating to these problems in a manner useful to engineers in understanding and designing solar processes.

Although solar energy may be considered a new or unconventional resource, it is used (and has been historically) for several applications. Solar evaporation of salt brines to recover the salts has been practiced for many centuries, as has solar drying of agricultural products. Solar water heating is a standard method of providing domestic hot water in parts of Australia, Israel, and Japan, and small but viable industries are based on the manufacture, sale, and installation of solar water heating equipment in these countries.

We restrict ourselves to considerations of thermal processes, in which solar radiation is absorbed by a surface and converted to heat. This heat is then stored and/or used directly in a subsequent operation--for water heating, building heating, operation of conversion devices such as absorption coolers, and other applications. We do not treat photovoltaic, photochemical, or photosynthetic processes. The emphasis is on solar energy for meeting thermal energy needs for buildings; the technology of other solar thermal processes, including conversion of solar energy to mechanical energy by thermal means, is basically the same as that presented here for heating and cooling. The basic techniques outlined for component and system analysis are applicable to other processes. We include only brief notes and references to solar ponds, solar power, and solar distillation which are intended to lead to the pertinent literature on these processes.

The development of subject matter in the book is based on 20 years of research at the Solar Energy Laboratory of the

University of Wisconsin. It is a direct outgrowth of the approach we are taking to our current solar energy studies, that
is, the development of models and simulation techniques that
can be used to improve understanding of how solar thermal processes work and how they can be designed and evaluated. Thus,
we present the material in forms that can readily be used in
computer simulations (although much of it can also be used as
the basis for hand calculations), in the belief that the availability of computing capability is so common that its use will
constitute a major method of designing solar energy processes
in the future.

The subject matter covered in this book can be considered
in four parts:

First, in Chapters 1 to 3, solar radiation, its origin, its
nature, and its availability are outlined. Solar radiation
data and its manipulation into forms useful in solar process
calculations are shown, as background for solar design calculations.

Second, in Chapters 4 to 6, we review some heat transfer
considerations that are important in solar energy. In particular, radiation heat transfer and properties of materials are
noted.

Third, in Chapters 7 to 10, we treat process and system
models. Flat-plate collectors are analysed in detail, as they
are basic to the solar processes of most interest. Focusing
collectors and energy storage are then treated, to establish
the principles and equations describing their thermal performance. Chapter 10 discusses system models and simulations, that
is, the procedures for solving the systems of component equations.

Fourth, in Chapters 11 to 14, we discuss a series of applications that have been the subject of analysis, experiment, and
application. Chapter 11 is on water heating, Chapter 12 on
space heating, and Chapter 13 on air conditioning, with each
adding new components and loads to systems treated in the preceding chapter. Chapter 14 notes some possible alternative
methods for solar heating and cooling that do not fit the development of the first three chapters of this part; this chapter
is intended to broaden the reader's perspectives on the range of
possible heating and cooling processes that can be considered.

Units have constituted a major problem. A variety of units
appear in the solar energy literature, and the reader of this
literature must be prepared to cope with the units as he finds
them. Under these circumstances, we have adopted the following
policy. General information and work done and published in
other units are usually shown in the units of the original authors. However, if the material is such that it constitutes a

source of data for engineering calculations, we have converted
it and present it in SI units. Original work of ours is in SI
units, and we recommend that the user work in these units (with
a time base of hours, rather than seconds). We use both kilo-
watts and kilojoules per hour for energy rates and heat transfer coefficients. The reader will find both of these units in
the literature and should use care in avoiding confusion between
them.

    We have included sample problems in appropriate parts of
the text, to illustrate methods of calculation and the kinds of
results to be expected. Additional problems are included in
Appendix A.

    In the preparation of these notes, we have drawn heavily
on a wide variety of sources. They are indicated in the reference lists of the individual chapters. These references are not
complete and do not contain a listing of all publications on the
subject, but they should be useful to the reader in obtaining
background on a particular topic in more detail than we present.
The bibliography at the end of this introduction includes a variety of general references in solar energy, such as proceedings
of symposia and conferences, and reference volumes on particular
topics. It also includes journals devoted to solar energy.
Particularly useful is the *Journal of Solar Energy*, published
by the International Solar Energy Society, which is a source of
a large volume of useful material in this field.

    This book is based on many sources of information, but we
have used the developments of several groups to an extent that
particular note is in order. They are: the MIT Solar Energy
Program and the work of H. C. Hottel and his colleagues, the
publications of A. Whillier (from MIT and subsequently from
McGill University), the work of G. O. G. Löf on space heating
and economics, the many publications of the Mechanical Engineering Division of CSIRO, Australia, and the papers of R. C. Jordan
and colleagues of the University of Minnesota.

    Many individuals have assisted us in the development of
this book. To list all of the contributors would be impossible.
The graduate students in our laboratory have been an invaluable
source of ideas, comments, and constructive criticism; some of
the material in the later chapters of the book is based on their
research, and references to their theses and papers are explicitly included. The help of students in our course on Solar Energy Technology over the past three years has also been of as-
sistance.

    It is also appropriate that we acknowledge the inspiration
of the late Farrington Daniels. He was instrumental in estab-
lishing the Solar Energy Laboratory at Wisconsin in 1954 and
provided much of the spark that kept interest in this field

alive during lean years. Although his interests were in solar
energy applications in developing countries, he was an effective
spokesman for solar energy research, and progress in this field
would not be as rapid as it is if he had not been active in it.

We acknowledge support of our research by the RANN program
of the National Science Foundation. Although the grant has not
directly supported the preparation of this book, we could not
have written it without the background and experience made possible by the grant.

We acknowledge with thanks the constructive criticisms and
contributions of our colleagues and students. In particular,
the manuscript is improved as a result of the helpful comments
by P. I. Cooper, D. J. Close, K. G. T. Hollands, G. O. G. Löf,
and S. L. Sargent.

We appreciate the cooperation of the many authors and pub-
lishers who have readily given us permission to use drawings or
other material from their work. We have attempted to reference
adequately all such sources.

Finally, we acknowledge the skill, diligence, patience and
good spirits of Mary Stampfli, who has typed the book.

Madison, Wisconsin   John A. Duffie
June 7, 1974   William A. Beckman

GENERAL PUBLICATIONS ON SOLAR ENERGY

Space Heating with Solar Energy, R. W. Hamilton, Ed., Cambridge,
Massachusetts Institute of Technology, 1954.

Solar Energy Research, F. Daniels and J. A. Duffie, Eds.,
Madison, University of Wisconsin Press, 1955.

Wind and Solar Energy, Proceedings of the New Delhi Symposium,
United Nations Education, Scientific, and Cultural Organization;
19 ave Kleber, Paris-16^e, France (1956). Printed by Imprimerie
Chaix, 20 rue Bergere, Paris-9^e, France.

Proceedings of the World Symposium on Applied Solar Energy,
Phoenix, Ariz., Stanford Research Inst., Menlo Park, Calif.
(1956).

New Sources of Energy and Economic Development, United Nations,
Department of Economics and Social Affairs, New York (1957),
(Sales #1957.II.B.1).

Transactions of the Conference on the Use of Solar Energy, The
Scientific Basis, Tucson, University of Arizona Press (1958).

Applications Thermiques de L'Energie Solaire Dans le Domaine de la Recherche et de L'Industrie, Proc. of Colloques Internationaux du Centre National de la Recherche Scientifique, Montlouis, June 23 to 28, 1958. Printed by Centre National de la Recherche Scientifique, 15 Quai Anatole-France, Paris (VIIe) (1961).

Introduction to the Utilization of Solar Energy, A. M. Zarem and D. D. Erway (Eds.), University of California Engr. and Sciences, Ext. Series, McGraw-Hill, 1963.

Proceedings of the UN Conference on New Sources of Energy, (1961, Rome), New York (1964). Volumes 4, 5 and 6. (UN sales numbers 63.I.38, 63.I.39 and 63.I.40.)

Direct Use of the Sun's Energy, F. Daniels, New Haven, Conn., Yale University Press (1964).

World Distribution of Solar Radiation, G. O. G. Löf, J. A. Duffie, and C. O. Smith. Report #21 published by the Engineering Experiment Station, Madison, University of Wisconsin, 1966.

Low Temperature Engineering Application of Solar Energy, R. C. Jordan, Ed. Published by the American Society of Heating, Refrigeration and Air Conditioning Engineers, New York, 1967.

Solar Energy in Developing Countries: Perspective and Prospects. A report of an ad hoc advisory panel to the National Academy of Science, Washington (March 1972).

Solar Energy as a National Energy Resource, NSF/NASA Solar Energy Panel Report, (1972) available from National Technical Information Service, 5285 Port Royal Rd., Springfield, Va. 22151, Report #PB221659.

Solar Energy for Man, B. J. Brinkworth, Compton, Chamberlayne, Salisburg, England, The Compton Press Ltd., 1972.

SOLAR ENERGY JOURNALS

Solar Energy: The Journal of Solar Energy Science and Technology, published by the International Solar Energy Society. (U. S. address: International Solar Energy Society, c/o Smithsonian Radiation Biology Lab., 12441 Parklawn Dr., Rockville, Maryland 20852.)

Cooperation Mediterraneenne pour L'Energie Solaire, Bulletin, published by COMPLES, 32 Cours Pierre-Puget, 13006 Marseilles, France.

Heliotechnology (Gelioteckhnika), a Russian language journal published in English translation by Allerton Press, New York.

CONTENTS

Solar Energy
Thermal Processes

1. EXTRATERRESTRIAL SOLAR RADIATION

The sun has structure and characteristics which determine the nature of the energy it radiates into space. This chapter notes the characteristics of this energy outside of the earth's atmosphere, and is the basis of the next chapter, which deals with the effects of the atmosphere in attenuating the radiation and the characteristics of the resulting energy resource available at the earth's surface. We are concerned here primarily with radiation in a wavelength range of 0.3 to 3.0 μm, that portion of the spectrum which includes most of the energy of solar radiation.

1.1 *THE SUN*

The sun is a sphere of intensely hot gaseous matter with a diameter of 1.39×10^6 km, and is, on the average, 1.5×10^8 km from the earth. As seen from the earth, the sun rotates on its axis about once every four weeks. However, it does not rotate as a solid body, the equator taking about 27 days and the polar regions about 30 days for each rotation.

The surface of the sun is at an effective temperature of about 5762°K*. The temperature in the central interior regions is variously estimated at 8×10^6 to 40×10^6°K, and the density about 80 to 100 times that of water. The sun is, in effect, a continuous fusion reactor with its constituent gases as the "containing vessel" retained by gravitational forces. The fusion reactions which have been suggested to supply the energy radiated by the sun have been several; the one considered the most important is a process in which hydrogen (i.e., four protons) combines to form helium (i.e., one helium nucleus); the mass of the helium nucleus is less than that of the four protons, mass having been lost in the reaction and converted to energy.

This energy is produced in the interior of the solar sphere, at temperatures of many millions of degrees. It must

*This leads to the same total energy as is received from the sun above the atmosphere. See Thekaekara (1974).

1

transfer out to the surface and then be radiated into space. A
succession of radiative and convective processes must occur,
with successive emission, absorption, and reradiation; the radiation in the sun's core must be in the x-ray and gamma ray
parts of the spectrum with the wavelengths of the radiation increasing as the temperature drops at larger radial distances.
     A schematic of the structure of the sun is shown in Figure
1.1.1. It is estimated that 90% of the energy is generated in
the region 0 to 0.23 R (where R = radius of the sun), which contains 40% of the mass of the sun. At a distance 0.7 R from the
center, the temperature has dropped to about 130,000°K and the
density has dropped to 0.07 g/cc; here convection processes begin to become important and the zone from 0.7 to 1.0 R is known
as the *convective zone*. Within this zone, the temperature
drops to about 5000°K and the density to about 10^{-8} g/cc.

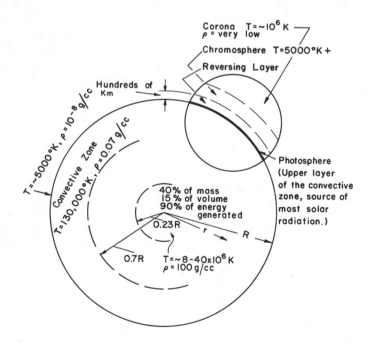

Figure 1.1.1 Schematic of the structure of the sun.

     The sun's surface appears to be composed of granules (ir-
regular convection cells), with dimensions of cells estimated

as being from 1000 to 3000 km and with cell lifetime of a few minutes. The upper layer of the convective zone is called the *photosphere*. The edge of the photosphere is sharply defined, even though it is of low density (about 10^{-4} that of air at sea level). It is essentially opaque, as the gases of which it is composed are strongly ionized and able to absorb and emit a continuous spectrum of radiation. The photosphere is the source of most solar radiation.

Outside of the photosphere is a more or less transparent solar atmosphere, which is observable during total solar eclipse or by instruments that occult the solar disk. Above the photosphere is a layer of cooler gases several hundred miles deep called the *reversing layer*. Outside of that is a layer referred to as the *chromosphere*, with a depth of about 10,000 km. This is a gaseous layer with temperature somewhat higher than that of the photosphere and with lower density. Still further out is the *corona*, of very low density and of very high (10^{6} °K) temperature.

This simplified picture of the sun, its physical structure, and its temperature and density gradients, will serve as a basis for appreciation that the sun does not, in fact, function as a blackbody radiator at a fixed temperature. Rather, the emitted solar radiation is the composite result of the several layers which emit and absorb radiation of various wavelengths. There is also some minor variability with time in spectral intensity distribution in the very short- and long-wave parts of the spectrum. For many purposes (e.g., thermal process), it is adequate to consider the sun as a blackbody radiator at about 5762°K; for other processes which are wavelength-dependent and where the spectral distribution is important (e.g., in photochemical or photovoltaic processes), more detailed considerations may be necessary. For further information on this subject see Thomas (1958), or Elson (1974).

1.2 THE SOLAR CONSTANT

Figure 1.2.1 shows schematically the geometry of the sun-earth relationships. The eccentricity of the earth's orbit is such that the distance between the sun and the earth varies by ±3%. At a distance of one astronomical unit, the mean earth-sun distance, the sun subtends an angle of 32'. The characteristics of the sun and its spatial relationship to the earth result in a nearly fixed intensity of solar radiation outside of the earth's atmosphere. *The solar constant*, I_{sc}, is the energy from the sun, per unit time, received on a unit area of surface perpendicular to the radiation, in space, at the earth's mean distance from the sun.

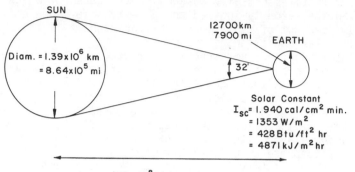

Figure 1.2.1 Schematic of sun-earth relationships (not to scale). The angle subtended by the sun at mean earth-sun distance is 32'.

Until recently, estimates of the solar constant had to be made from ground-based measurements of solar radiation after it had been transmitted through the atmosphere, and thus in part absorbed and scattered by components of the atmosphere. Extrapolations from the terrestrial measurements, which were made from high mountains, had to be based on estimates of atmospheric transmission in various portions of the solar spectrum. Pioneering studies were done by C. G. Abbot and his colleagues at the Smithsonian Institution. These studies and later measurements from rockets were summarized by Johnson (1954); Abbot's value of the solar constant of 1322 W/m^2, was revised upward by Johnson to 1395 W/m^2.

More recently, the availability of very high altitude aircraft, balloons, and spacecraft has permitted direct measurements of solar intensity outside most or all of the earth's atmosphere. These measurements have been reviewed and summarized, and a new standard value of the solar constant proposed by Thekaekara and Drummond (1971), as 1353 W/m^2 (1.940 cal/cm^2 min, 428 Btu/ft^2 hr, or 4871 kJ/m^2 hr).

1.3 *SPECTRAL DISTRIBUTION OF EXTRATERRESTRIAL RADIATION*

In addition to the total energy (i.e., solar constant) in the solar spectrum, it is useful to know the spectral distribution of this radiation. The accepted value of the solar constant has changed primarily because of improved knowledge of extra-terrestrial radiation in particular portions of the spectrum. A new standard spectral irradiance curve has been compiled, based on high altitude and space measurements. The new proposed standard from NASA (1971) is shown in Figure 1.3.1. The aver-aged energy over small bandwidths centered at wavelength λ and the integrated energy for wavelengths less than λ for the new proposed standard curve are indicated in Table 1.3.1.

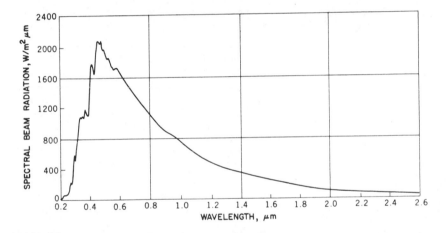

Figure 1.3.1 The NASA (1971) Standard spectral irradiance at the mean sun-earth distance and a solar constant of 1353 W/m².

Example 1.3.1 Calculate the fraction of the extraterres-trial solar radiation and the amount of that radiation in the ultraviolet ($\lambda < 0.38$ μm), the visible (0.38 μm $< \lambda < 0.78$ μm), and the infrared ($\lambda > 0.78$ μm) portions of the spectrum.

From Table 1.3.1, the fractions D_λ corresponding to wave-lengths of 0.38 and 0.78 μm are 0.0700 and 0.5429 (inter-polated). Thus, the fraction in the ultraviolet is 0.0700, the fraction in the visible range is (0.5429 - 0.0700)

= 0.4729, and the fraction in the infrared is (1.0 -
0.5429) = 0.4571. Applying these fractions to a solar
constant of 1353 W/m^2 and tabulating the results, we get

Wavelength range, μm	0 - 0.38	0.38 - 0.78	0.78 - ∞
Fraction in range	0.0700	0.4729	0.4571
Energy in range, W/m^2	95	640	618

TABLE 1.3.1 Extraterrestrial Solar Irradiance (Solar Constant
= 1353 W/m^2).*

λ	E_λ	D_λ	λ	E_λ	D_λ	λ	E_λ	D_λ
0.115	.007	1×10^{-4}	0.43	1639	12.47	0.90	891	63.37
0.14	.03	5×10^{-4}	0.44	1810	13.73	1.00	748	69.49
0.16	.23	6×10^{-4}	0.45	2006	15.14	1.2	485	78.40
0.18	1.25	1.6×10^{-3}	0.46	2066	16.65	1.4	337	84.33
0.20	10.7	8.1×10^{-3}	0.47	2033	18.17	1.6	245	88.61
0.22	57.5	0.05	0.48	2074	19.68	1.8	159	91.59
0.23	66.7	0.10	0.49	1950	21.15	2.0	103	93.49
0.24	63.0	0.14	0.50	1942	22.60	2.2	79	94.83
0.25	70.9	0.19	0.51	1882	24.01	2.4	62	95.86
0.26	130	0.27	0.52	1833	25.38	2.6	48	96.67
0.27	232	0.41	0.53	1842	26.74	2.8	39	97.31
0.28	222	0.56	0.54	1783	28.08	3.0	31	97.83
0.29	482	0.81	0.55	1725	29.38	3.2	22.6	98.22
0.30	514	1.21	0.56	1695	30.65	3.4	16.6	98.50
0.31	689	1.66	0.57	1712	31.91	3.6	13.5	98.72
0.32	830	2.22	0.58	1715	33.18	3.8	11.1	98.91
0.33	1059	2.93	0.59	1700	34.44	4.0	9.5	99.06
0.34	1074	3.72	0.60	1666	35.68	4.5	5.9	99.34
0.35	1093	4.52	0.62	1602	38.10	5.0	3.8	99.51
0.36	1068	5.32	0.64	1544	40.42	6.0	1.8	99.72
0.37	1181	6.15	0.66	1486	42.66	7.0	1.0	99.82
0.38	1120	7.00	0.68	1427	44.81	8.0	.59	99.88
0.39	1098	7.82	0.70	1369	46.88	10.0	.24	99.94
0.40	1429	8.73	0.72	1314	48.86	15.0	4.8×10^{-2}	99.98
0.41	1751	9.92	0.75	1235	51.69	20.0	1.5×10^{-2}	99.99
0.42	1747	11.22	0.80	1109	56.02	50.0	3.9×10^{-4}	100.00

*E_λ is solar spectral irradiance averaged over small bandwidth
centered at λ, in W/m^2 μm; D_λ is percentage of the solar constant associated with wavelengths shorter than λ. From
Thekaekara (1974)

REFERENCES

Elson, B. M., Aviation Week and Space Technology, *63*, (January 14, 1974). "Theoretical Picture of Sun Still Evolving."

Johnson, F. S., J. Meteorol., *11*, 431 (1954). "The Solar Constant."

NASA SP-8005, National Aeronautics and Space Administration, May (1971). "Solar Electromagnetic Radiation."

Thekaekara, M. P. and Drummond, A. J., Nat. Phys. Sci., *229*, 6 (1971). "Standard Values for the Solar Constant and Its Spectral Components."

Thekaekara, M. P., Supplement to Proc. 20th Annual Meeting of Inst. for Environmental Sci., 21 (1974). "Data on Incident Solar Radiation."

Thomas, R. N., Transactions of the Conference on Use of Solar Energy, *1*, 1, University of Arizona Press, 1958. "The Features of the Solar Spectrum as Imposed by the Physics of the Sun."

2. SOLAR RADIATION AT EARTH'S SURFACE

Chapter 1 noted the intensity and spectral distribution of solar radiation outside of the atmosphere at earth-sun distance.
Here we consider factors affecting radiation intensity at the
ground, that is, atmospheric attenuation and receiving-surface
orientation. In Chapter 3 we will consider measurement of solar radiation, the data that are available, and methods of conversion of those data to the forms useful in predicting and analyzing the performance of solar processes.

In general, it is not practical to start from knowledge of
extraterrestrial radiation and predict the intensity and spectral distribution to be expected on the ground. Adequate meteorological data for such calculations are seldom available, and
recourse is usually made to measurements. However, an understanding of atmospheric attenuation and effects of orientation
of receiving surface is useful in understanding and using solar
radiation data.

2.1 *DEFINITIONS*

Several definitions will be useful in understanding this chapter:

Beam Radiation is that solar radiation received from the sun
without change of direction.

Diffuse Radiation is that solar radiation received from the sun
after its direction has been changed by reflection and scattering by the atmosphere.

Air Mass, m, is the path length of radiation through the atmosphere, considering the vertical path at sea level as unity.
Thus at sea level, $m = 1$ when the sun is at the zenith (i.e.,
directly overhead), and $m = 2$ for a zenith angle (θ_z, the angle
subtended by the zenith and the line of sight to the sun) of
60°. For all but very high zenith angles ($m > 3$, where the
earth's curvature becomes significant),

$$m = \sec \theta_z \qquad (2.1.1)$$

8

2.2 VARIATION OF EXTRATERRESTRIAL RADIATION

It has been determined from analyses of radiation data that the variations in total radiation emitted by the sun are probably less than ±1.5% [see Moon (1940) or Thekaekara (1965)]. For purposes of thermal processes that use energy in large fractions of the total solar spectrum, and where the transmittance of the atmosphere is a major uncertainty, the emission of energy by the sun can be considered as constant.

Variations in earth-sun distance do, however, lead to variations of extraterrestrial radiation flux in the range of ±3%. The dependence of extraterrestrial radiation with time of year is shown in Figure 2.2.1.

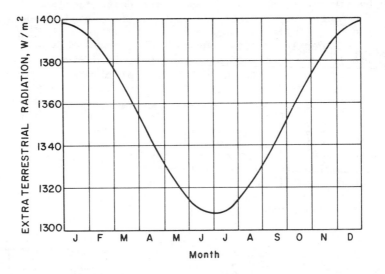

Figure 2.2.1 Variation of the extraterrestrial solar radiation with time of the year.

2.3 ATTENUATION OF BEAM RADIATION

Radiation at normal incidence received at the surface of the earth from the sun is subject to variations due to (1) variations in distance from earth to sun, (2) variations in atmospheric scattering by air molecules, water vapor, and dust, and (3) variations in atmospheric absorption by O_2, O_3, H_2O, and CO_2.

The normal solar radiation incident on the earth's atmosphere has a spectral distribution, indicated by Figure 1.3.1. The x-rays and other very short-wave radiations of the solar spectrum are absorbed high in the ionosphere by nitrogen, oxygen, and other atmospheric components; most of the ultraviolet is absorbed by ozone. At wavelengths longer than 2.5 μm, a combination of low extraterrestrial radiation and strong absorption by CO_2 and H_2O means that very little energy reaches the ground. Thus, from the viewpoint of terrestrial applications of solar energy, only radiation of wavelengths between 0.29 and 2.5 μm need be considered. This solar radiation is transmitted through the atmosphere, undergoing variations due to scattering and absorption.

Scattering, which results in attenuation of the beam radiation by air molecules, water vapor, and dust, has been the subject of a number of studies, and approximate methods have been developed to estimate the magnitude of the effect. This question and the pertinent literature up to 1940 are treated by Moon (1940), from whom much of this survey is derived. A more recent discussion, including the effects of clouds, is given by Fritz (1958). Thekaekara (1974) summarizes the present state of knowledge and includes an extensive bibliography.

Air molecules are very small compared to the wavelengths of radiation significant in the solar energy spectrum. Scattering of this radiation by molecules occurs in accordance with the theory of Rayleigh, which indicates that the scattering coefficient would vary approximately as λ^{-4}, where λ is the wavelength of the radiation. This has been experimentally verified, and the monochromatic transmission factor associated with atmospheric scattering can be written

$$\tau_{a\lambda} = 10^{-0.00389\lambda^{-4}} \qquad (2.3.1)$$

where λ is in micrometers, $m = 1$, and the barometric pressure is 760 mm.

Dust scattering from particles that are much larger than air molecules and that vary in size and concentration from location to location, height, and from time to time, is more difficult to assess. Moon developed a transmission factor of form similar to that for air molecules:

$$\tau_{d\lambda} = 10^{-0.0353\lambda^{-0.75}} \qquad (2.3.2)$$

where $m = 1$ and the average concentration of dust particles is $800/cm^3$ at the ground.

Water vapor scattering for zenith sun and 20 mm of precipitable water (the amount of water vapor in the air column above the observer) can be written as

$$\tau_{w\lambda} = 10^{-0.0075\lambda^{-2}} \qquad (2.3.3)$$

The total effect of scattering* on the beam radiation can be written as an approximation:

$$\tau_{\lambda(s)} = \left[(\tau_{a\lambda})^{p/760} \ (\tau_{d\lambda})^{d/800} \ (\tau_{w\lambda})^{w/20} \right]^{m} \qquad (2.3.4)$$

where $\tau_{\lambda(s)}$ = monochromatic atmospheric transmittance for beam radiation, considering scattering only, at wavelength λ;

p = total pressure, mm;
d = dust particle concentration, at the ground, particles/cm^3;
w = depth of precipitable water, mm;
m = air mass.

Note that the atmospheric pressure, air mass, concentration of dust, and amount of water vapor enter in an exponential manner in determining the attenuation of the beam radiation. The first two of these parameters are readily determined. The values of d and w are generally not known, except for estimates based on surface measurements.

Absorption of radiation in the atmosphere in the solar energy spectrum is due largely to ozone in the ultraviolet, and water vapor in bands in the infrared. There is almost complete absorption of short-wave radiation below 0.29 μm, and for typical values of ozone in the atmosphere, transmittance values due to ozone are shown in Table 2.3.1.

Water vapor absorbs strongly in the infrared bands, as illustrated in Figure 2.3.1, which shows variation of transmission due to water vapor as a function of λ. Beyond 2.3 μm, the transmission of the atmosphere is very low due to absorption by H_2O and CO_2, the energy in the extraterrestrial solar energy spectrum is less than 5% of the total solar spectrum, and energy received at the ground is small.

The transmittances for absorption are to be combined in the same manner as those for scattering, and the resulting

*According to the Rayleigh theory of scattering, the shorter wavelengths are scattered most and, hence, diffuse radiation will tend to be at shorter wavelengths.

monochromatic transmittance for beam radiation may be written as

$$\tau_\lambda = \tau_{\lambda(s)}\tau_{\lambda(abs)} = \tau_{\lambda(s)}\tau_{o\lambda}\tau_{w\lambda} \qquad (2.3.5)$$

Note that at least one of the absorption transmission factors $\tau_{o\lambda}$ (for ozone) or $\tau_{w\lambda}$ (for water) will be unity, as the wavelength ranges of absorption by O_3 and H_2O do not overlap.

TABLE 2.3.1 Ultraviolet Transmission of the Atmosphere, for a Layer of Ozone 2.5 mm Deep at NTP [Fritz (1958)].*

λ, μm	$\tau_{o\lambda}$
0.29	0
0.30	0.10
0.31	0.50
0.33	0.90
0.35	1.00

*Note. There is also a weak absorption band near 0.6 μm, where $\tau_{o\lambda}$ is > 0.95.

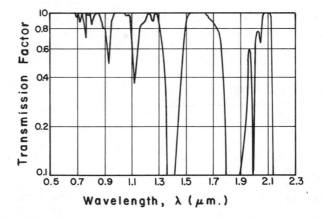

Figure 2.3.1 Transmission factor associated with absorption by water vapor [from Fowle (1915)].

Calculations of this type have been summarized by Moon, (1940) who used the best data available on the spectral distribution of solar radiation outside the atmosphere and on the various transmission factors for typical atmosphere to generate a series of "proposed standard curves" for beam solar radiation as a function of wavelength at air masses of 0 to 5. He also tabulates the intensities for a standard atmosphere (p = 760 mm, w = 20 mm, d = 300/cm^3, ozone = 2.8 mm) for various wavelengths for m = 2, and proposed that it be used as a standard spectral distribution curve for beam solar radiation for locations near sea level.

More recent studies of Johnson and of Thekaekara and Drummond, noted in Chapter 1, have indicated that Moon's intensities in the short-wave end of the spectrum for m = 0 are too low. Moon used a solar constant of 1322 W/m^2 in contrast to the now accepted value of 1353 W/m^2. Thekaekara (1974) presents new spectral distribution curves for beam radiation, based on the extraterrestrial distribution shown in Figure 1.3.1, for very clear and relatively clear atmospheres, for air masses of 1, 4, 7 and 10. Figure 2.3.2, from Thekaekara, shows the extraterrestrial distribution, its comparison with a normalized blackbody distribution curve for a source temperature of 5762°K, and for the very clear atmosphere the air mass one solar spectrum without and with molecular absorption. The latter distribution (one of the set of new distribution curves noted above) shows absorption bands due to O_3, H_2O, CO_2, and O_2.

2.4 *DIFFUSE RADIATION AT THE GROUND*

As noted above, components of the atmosphere scatter a portion of the solar radiation, and some of this scattered radiation reaches the ground. Thus there is always some diffuse radiation, even in periods of very clear skies. Particles of water and solids in clouds scatter radiation, and in periods of heavy clouds, all of the radiation reaching the ground will be diffuse. A useful discussion of diffuse radiation (scattered by clear atmosphere or by clouds) is provided by Fritz (1958).

As a practical matter, adequate atmospheric data are not available on which to base a computation of intensity, spectral distribution, or directional distribution of diffuse radiation at the ground. The directional distribution of diffuse radiation is discussed in section 3.7.

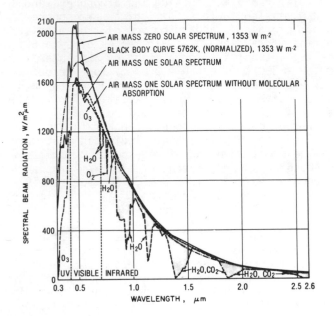

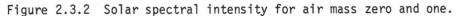

Figure 2.3.2   Solar spectral intensity for air mass zero and one.

2.5 DIRECTION OF BEAM RADIATION

The geometric relationships between a plane of any particular
orientation relative to the earth at any time (whether that
plane is fixed or moving relative to the earth) and the incoming beam solar radiation, that is, the position of the sun relative to that plane, can be described in terms of several angles.
These angles, and the relationships between them [from Benford
and Bock (1939)], are:

ϕ = latitude (north positive);
δ = declination (i.e., the angular position of the sun at
     solar noon with respect to the plane of the equator)
     (north positive);
s = the angle between the horizontal and the plane (i.e.,
     the slope);
γ = the surface azimuth angle, that is, the deviation of
     the normal to the surface from the local meridian, the

zero point being due south, east positive, and west
negative;

ω = hour angle, solar noon being zero, and each hour equal-
ing 15° of longitude with mornings positive and after-
noons negative (e.g., ω = +15 for 11:00, and ω = -37.5
for 14:30);

θ = the angle of incidence of beam radiation, the angle
being measured between the beam and the normal to the
plane.

The declination, δ, can be found from the approximate equation
of Cooper (1969):

$$\delta = 23.45 \sin \left[360 \ \frac{284 + n}{365} \right] \qquad (2.5.1)$$

where n is the day of the year.* The relation between θ and the
other angles is given by

$$\cos \theta = \sin \delta \sin \phi \cos s - \sin \delta \cos \phi \sin s \cos \gamma$$

$$+ \cos \delta \cos \phi \cos s \cos \omega$$

$$+ \cos \delta \sin \phi \sin s \cos \gamma \cos \omega$$

$$+ \cos \delta \sin s \sin \gamma \sin \omega$$

$$(2.5.2)$$

Example 2.5.1 Calculate the angle of incidence of beam
radiation on a surface located at Madison, Wisconsin at
14:30 on February 15, if the surface is tilted 45° from
the horizontal and is pointed 15° west of south.

Under these conditions, the declination is -14°, the hour
angle is -37.5°, and the surface azimuth angle is -15°.
Using the slope of 45° and Madison's latitude of 43°N, Eq.
(2.5.2) is

$$\cos \theta = \sin (-14) \sin 43 \cos 45$$

$$- \sin (-14) \cos 43 \sin 45 \cos (-15)$$

$$+ \cos (-14) \cos 43 \cos 45 \cos (-37.5)$$

$$+ \cos (-14) \sin 43 \sin 45 \cos (-15) \cos (-37.5)$$

*Declination can also be conveniently determined from charts
such as Figure 3.4.2.

$$+ \cos (-14) \sin 45 \sin (-15) \sin (-37.5)$$

$$\cos \theta = -0.1167 + 0.1208 + 0.3981 + 0.3586 + 0.1081$$

$$\cos \theta = 0.8689$$

$$\theta = 30°$$

Additional angles may also be defined. The most frequently used are

θ_z = zenith angle, the angle between the beam from the sun and the vertical;

α = solar altitude, the angle between the beam from sun and the horizontal, equal to $(90° - \theta_z)$.

(In architectural and illumination practice, other angles are defined, such as the profile angle. Care must be exercised in the use of any source of information on these angles so that their authors' definitions and sign conventions are understood.)
    In many cases, the equation relating these angles is simplified. For example, for fixed flat plate collectors which face the equator, the last term drops out. For vertical surfaces, s = 90°, and the first and third terms drop out. For horizontal surfaces where s = 0°, only the first and third terms remain, and the angle of incidence (i.e., the zenith angle of the sun) is

$$\cos \theta_z = \sin \delta \sin \phi + \cos \delta \cos \phi \cos \omega \qquad (2.5.3)$$

Example 2.5.2 Calculate the zenith angle of the sun at Madison at 14:30 on February 15.

For this situation, Eq. (2.5.3) is

$$\cos \theta_z = \sin (-14) \sin 43 + \cos (-14) \cos 43 \cos (-37.5)$$

$$\theta_z = 0.3980$$

$$\theta_z = 66°$$

Useful relationships for the angle of incidence on surfaces sloped to the north or south can be derived from the fact that surfaces with slope s to the north or south have the same angular relationship to beam radiation as a horizontal surface at an artificial latitude of $(\phi - s)$. The relationship is shown in Figure 2.5.1. Then, modifying Eq. (2.5.3),

$$\cos\ \theta_T = \cos\ (\phi - s)\ \cos\ \delta\ \cos\ \omega + \sin\ (\phi - s)\ \sin\ \delta$$

$$(2.5.4)$$

Note that the slope, s, is measured from the horizontal to the plane of the surface in question, and is positive when slope is toward the south.

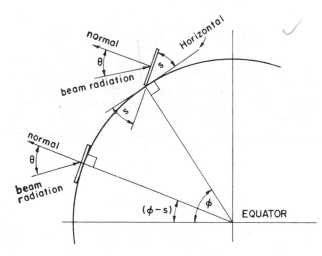

Figure 2.5.1   Section of earth showing definitions of s, θ, ϕ, and $(\phi - s)$.

Equation (2.5.3) can be solved for the sunrise hour angle, ω_s, when $\theta_z = 90°$:

$$\cos\ \omega_s = -\ \frac{\sin\ \phi\ \sin\ \delta}{\cos\ \phi\ \cos\ \delta}$$

$$\cos\ \omega_s = -\ \tan\ \phi\ \tan\ \delta$$

$$(2.5.5)$$

It also follows that the day length is given by

$$T_d = \frac{2}{15}\ \cos^{-1}(-\tan\ \phi\ \tan\ \delta) \qquad (2.5.6)$$

Solar azimuth and altitude angles are tabulated as functions of latitude, declination, and hour angle by the U.S. Hydrographic

Office (1940). Information on the position of the sun in the
sky is also available with less precision but easier access in
various types of charts. Examples of these are the Sun Angle
Calculator (1951) and diagrams in a paper by Hand (1948).
(Note that care is necessary in interpreting information from
sources such as these, since definitions of angles may vary
from those used here.) Brooks and Miller (1963) also present
a useful discussion of these geometrical relationships.

2.6 *DIRECTION OF DIFFUSE RADIATION*

Section 2.5 dealt with the direction of beam radiation, which
is a geometrical problem. The problem of coping with diffuse
radiation is a more difficult one, as the radiation (by definition) comes from parts of the sky dome other than the sun. The
distribution of diffuse radiation over the sky is highly variable depending on atmospheric conditions and will be discussed
in Chapter 3. Some measurements are available, for example as
cited by Fritz (1958).

2.7 *SOLAR TIME AND THE EQUATION OF TIME*

It is to be noted that the time specified in all of the sun
angle relationships is solar time, which does not coincide with
local clock time. It is necessary to convert standard time to
solar time by applying two corrections. First, there is a constant correction for any difference in longitude between the
location and the meridian on which the local standard time is
based (e.g., 90°W for the central time zone).* The second correction is from the *equation of time* which takes into account
the various perturbations in the earth's orbit and rate of rotation which affect the time the sun appears to cross the observer's meridian. This correction is obtained from published
charts. Thus, solar time can be obtained from standard time by
this relation:

$$\text{solar time} = \text{standard time} + E + 4\,(L_{st} - L_{loc}) \quad (2.7.1)$$

where E = the equation of time, from Figure 2.7.1, in minutes,
L_{st} = the standard meridian for the local time zone, and L_{loc}
= the longitude of the location in question, in degrees west.

*Note: Standard meridians for continental U.S. time zones are:
eastern, 75°W; central, 90°W; mountain, 105°W; and pacific, 120°W.

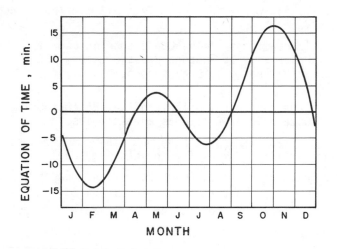

Figure 2.7.1 The equation of time, E, in minutes, as a function of time of year.

Example 2.7.1 What is solar time corresponding to 10:30 a.m. central standard time on February 2? In Madison the equation is

solar time = standard time + E + 4(90 - 89.38)

= standard time + E + 2.48

Note that the sun takes 4 min to traverse 1° of longitude, and that the units of the last term are therefore minutes. On February 2, E is -13.5 min, so the correction to standard time to get solar time is -11 min. Thus 10:30 a.m. central standard time is 10:19 a.m. solar time.

REFERENCES

Benford, F. and Bock, J. E., Trans. Am. Illum. Eng. Soc., 34, 200 (1939). "A Time Analysis of Sunshine."

Brooks, F. A. and Miller, W., Introduction to the Utilization of Solar Energy, Zarem, A. M. and Erway, D. D., Eds., New York, McGraw-Hill, 1963.

Cooper, P. I., Solar Energy, 12, 3 (1969). "The Absorption of Solar Radiation in Solar Stills."

Fowle, F. E., Astrophys. J., *42*, 379 (1915). "The Transparency of Aqueous Vapor."

Fritz, S., Transactions of the Conference on Use of Solar Energy, *1*, 17, University of Arizona Press, Tucson, 1958. "Transmission of Solar Energy Through the Earth's Clear and Cloudy Atmosphere."

Hand, I. F., Heating and Ventilating, *45*, 86 (October 1948). "Charts to Obtain Solar Altitudes and Azimuths."

Moon, P., J. Franklin Inst., *230*, 583 (1940). "Proposed Standard Solar Radiation Curves for Engineering Use."

Sun Angle Calculator, Libby-Owens-Ford Glass Company (1951).

Thekaekara, M. P., Solar Energy, *9*, 7 (1965). "The Solar Constant and Spectral Distribution of Solar Radiant Flux."

Thekaekara, M. P., Supplement to the Proc. 20th Annual Meeting of Inst. for Environmental Sci., 21 (1974). "Data on Incident Solar Energy."

U.S. Hydrographic Office Publication No. 214 (1940). "Tables of Computed Altitude and Azimuth."

3. SOLAR RADIATION:
MEASUREMENTS, DATA, AND ESTIMATION

It is generally not practical to base predictions of solar radiation availability on attenuation of extraterrestrial radiation, because the meteorological information necessary to make the calculations is seldom available. Instead, measurements of radiation from the location in question or from a nearby similar location can be used in solar process design, or estimates of solar radiation can be made from related meteorological data. In this chapter, we discuss methods of solar radiation measurement, the nature of the data that are available, and the manipulation of those data to forms useful in estimating the energy available to a solar process.

There are several approaches to the use of solar radiation data. One is to use average solar energy available, for example, for a month, to estimate the average performance of a process. This is unsatisfactory because the performance of many processes is not linear with solar radiation and the use of averages in these cases may lead to serious errors. A second is to use past hourly or daily data for the location in question, estimate what the performance of a process would have been under those past conditions, and on this basis project future performance. The latter is the general approach to process simulation which will be developed in later chapters. A third set of methods is to reduce the radiation data to more manageable form by statistical methods, and use the resulting time distributions in process performance predictions.*

3.1 DEFINITIONS

Figure 3.1.1 shows schematically the primary radiation fluxes on a surface at or near the ground; these fluxes are important in connection with solar thermal processes. For this purpose

*A discussion of one of these methods, a statistical treatment of radiation availability which is based in part on collector performance, is included in Chapter 7.

21

radiation is considered in two wavelength ranges:

1. *Solar, or short-wave radiation*: Radiation originating from the sun, at a source temperature of about 6000°K, and in the wavelength range of 0.3 to 3.0 μm.

2. *Long-wave radiation*: Radiation originating from sources at temperatures near ordinary ambient temperatures, and thus substantially all at wavelengths greater than 3 μm.

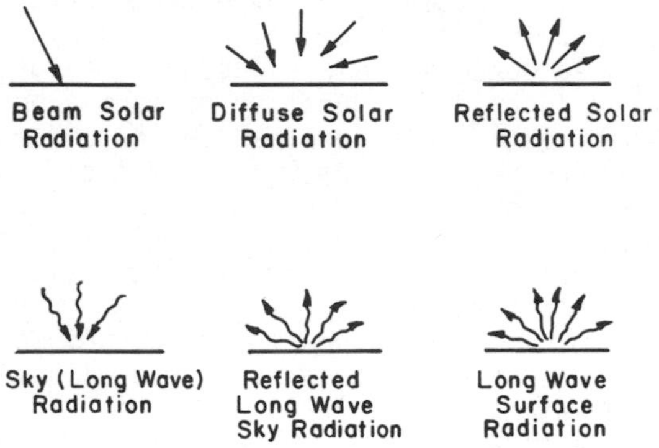

| Beam Solar Radiation | Diffuse Solar Radiation | Reflected Solar Radiation |
| Sky (Long Wave) Radiation | Reflected Long Wave Sky Radiation | Long Wave Surface Radiation |

Figure 3.1.1 The primary radiant energy fluxes of importance in solar thermal processes. Short wave solar radiation is shown by →. Long wave radiation is shown by ⤳.

The meteorological literature, in which radiation instruments and data are described and reported, uses several terms which are of interest to the solar process engineer:

1. *Pyrheliometer*: An instrument using a collimated detector for measuring solar radiation from a small portion of the sky including the sun (i.e., beam radiation) at normal incidence.

2. *Pyranometer*: An instrument for measuring total hemispherical solar (beam + diffuse) radiation, usually on a horizontal surface. If shaded from the beam radiation by a shade ring, it measures diffuse radiation.

3. *International Pyrheliometric Scale* (1956): Two standard reference pyrheliometers have been in use for many years (the Abbot, or Smithsonian instrument, and the Angstrom instrument). The scales defined by these instruments did not agree, and in 1956 a new "International Pyrheliometric Scale 1956" was established which applied corrections to previously obtained

measurements for each of these (-2.0% to the Smithsonian scale, +1.5% to the Angstrom scale). All instruments manufactured and calibrated since 1956 use the new scale. See Thekaekara (1965) for a discussion of pyrheliometric scales.

In addition, the term *solarimeter* and *actinometer* are encountered; they can generally be interpreted to be the same as pyranometer.

3.2 RADIATION MEASUREMENTS, INSTRUMENTS

Solar radiation measurements are most often made of total (beam and diffuse) radiation, in energy per unit time per unit area, on a horizontal surface, by pyranometer. Measurements are also made of beam radiation, by pyrheliometer, which respond to solar radiation received from a very small portion of the circumsolar sky. Instruments for these measurements convert radiation to some other form of energy and provide a measure of the energy flux produced by the radiation. Several brief reviews of solar radiation instruments are available, for example, by Morikofer (1958), by Drummond (1964), and by Yellott (1967).

The most common pyranometers in use in the United States today are based on detection of the difference between the temperature of black surfaces (which absorb most solar radiation), and white surfaces (which reflect most solar radiation) by thermopiles. Properly protected from wind, and compensated for changes in ambient temperature, the thermopiles give millivolt signals that can be readily detected, recorded, and integrated over time.

In this country, the *Eppley pyranometer* is based on this principle and has become the most common instrument in use by weather bureau stations. It uses concentric silver rings, 0.25 mm thick, appropriately coated black and white, with either 10 or 50 thermocouple junctions to detect temperature differences between the coated rings. Later models use wedges arranged in a circular pattern, with alternate black and white coatings. The disks or wedges are enclosed in a hemispherical glass cover. Its performance has been carefully studied by MacDonald (1951). Similar instruments are manufactured in Europe under the name *Kipp*.

The Eppley pyranometers, and similar instruments, are calibrated in a horizontal position. In general use, without frequent calibration the data from them are probably no better than ±5%. With frequent calibration against standard instruments, the data may be as good as ±2%. Calibrations of these instruments will vary to some degree if the instrument is in-clined to measure radiation on other than a horizontal surface.

The *Moll-Gorczynski solarimeter* is a pyranometer based on thermopiles having hot junctions exposed to solar radiation and cold junctions shaded from the radiation. Another type of pyra-nometer, the *Robitsch*, is based on differential expansion of bi-metal elements exposed to solar radiation; this has the advan-tage, where no power is available, of the possibility of direct mechanical linkage to a spring-driven recording drum. Thermal expansion radiometers are widely used in isolated stations, and while they are not precise, they are a major source of the solar radiation data that are available outside of Europe and North America.

Pyranometers have also been based on photovoltaic (solar cell) detectors, (e.g., the *Yellott solarimeter*). Silicon cells are the most common for solar energy, although cadmium sulfide and selenium cells have been used, for example, for measurements of visible light in photography. Silicon solar cells have the property that their light current (approximately equal to the short-circuit current at normal radiation levels) is a linear function of the incident solar radiation. They have the dis-advantage that the spectral response is not linear, so instru-ment calibration is a function of the spectral distribution of the incident radiation. Also, the calibration varies with the angle of incidence of the radiation.

Three pyrheliometers have been in widespread use to measure normal incidence beam radiation: the *Angstrom pyrheliometer*, the *Abbot water flow pyrheliometer*, and the *Abbot silver disc pyrheliometer*. These instruments provide primary and secondary standards of solar radiation measurements.

In addition to radiation measurements, there are in wide-spread use instruments for recording the duration of "bright sunshine." The standard U.S. Weather Bureau instrument uses a pair of photocells, one of which is shaded from beam radiation. When radiation is all diffuse, the cells indicate nearly equal radiation levels; when beam radiation is incident on the in-strument, the cell exposed to that beam radiation indicates higher intensity than the shaded cell. The duration of a criti-cal intensity difference detected by the two cells is a measure of duration of "bright sunshine." An older instrument, the *Campbell-Stokes recorder* uses a spherical lens that produces an image of the sun on a treated paper. The paper is burned when-ever the beam radiation is above a critical level. The lengths of the burned portions of the paper provide an index of the duration of sunshine.

The *Bellani distillation pyranometer* uses flat or spherical containers of alcohol which are connected to calibrated

condenser-receiving tubes. The quantity of alcohol condensed
is a measure of the integrated solar energy on the spherical or
flat receiver.

Most radiation data available are for horizontal surfaces,
include both direct and diffuse radiation, and were measured
with thermopile instruments (or, in some cases, bimetallic detectors). Most of these instruments provide radiation versus
time records and do not themselves provide a means of integrating the records. The data are usually recorded in a form similar to that shown on Figures 3.2.1 and 3.2.2 by recording po-
tentiometers and are integrated graphically or electronically.

There are in the United States about 88 stations recording
solar radiation and reporting it as langleys received on a horizontal surface per day. (1 langley = 1 cal/cm^2 = 4.186 J/cm^2.)
Outside of the United States there are well-developed networks
in some countries, but in many areas of potential solar applications, data are lacking. The situation is improving, and many
new stations have been reporting data during and since the
International Geophysical Year.

A major source of information on solar radiation and related meteorological data in the United States is the Environmental Data Service, National Climatic Center, National Oceanic
and Atmospheric Administration, Asheville, N.C. 28801. Data are
available on tape or punch cards for daily and (for some 40
stations) hourly solar radiation, for varying periods starting
from 1952. A few stations also measure beam radiation on clear
days. Figure 3.2.3 shows the locations of U.S. stations reporting daily solar radiation values, and Figure 3.2.4 show stations
reporting hourly values.

There are many hundreds of stations throughout the world
where data are available on the duration of sunshine, usually
recorded with instruments of the Campbell-Stokes type. The
problem of the relationship of sunshine hours to radiation energy received will be discussed below.

3.3 SOLAR RADIATION DATA

Solar radiation data are available in several forms, and should
include the following information:

   1. Whether they are instantaneous measurements or values
   integrated over some period of time (usually hour or day);
   2. The time or time period of the measurements;
   3. Whether the measurements are of beam, diffuse or total
 • radiation, and the instruments used;
   4. The receiving-surface orientation (usually horizontal,

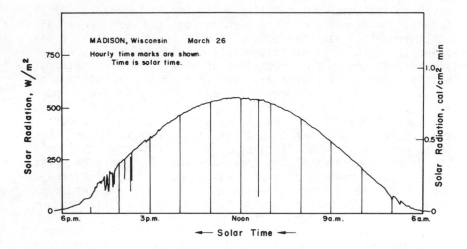

Figure 3.2.1 Total (beam + diffuse) solar radiation on a horizontal surface versus time for a clear day.

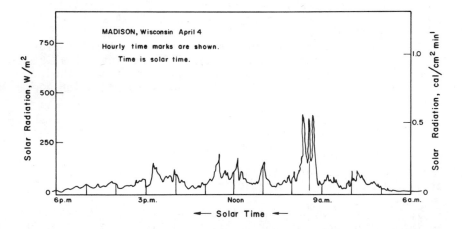

Figure 3.2.2 Total (beam + diffuse) solar radiation on a horizontal surface versus time for a cloudy day.

26

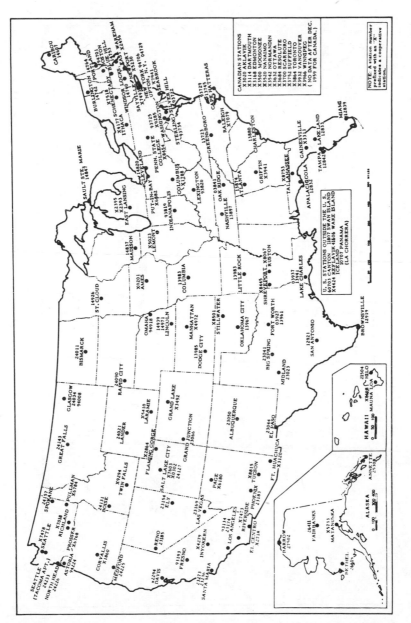

Figure 3.2.3 United States Weather Bureau stations measuring daily solar radiation as of 1967.

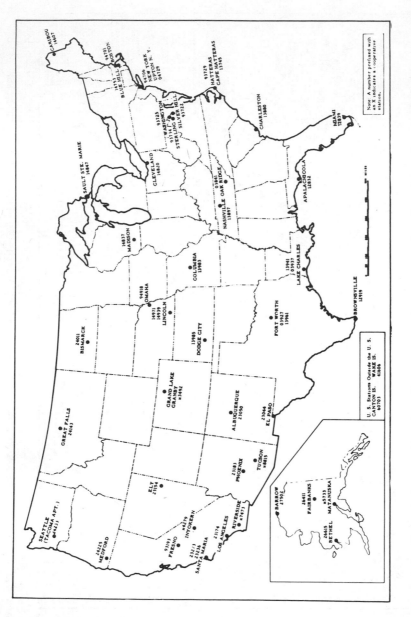

28

Figure 3.2.4 United States Weather Bureau stations measuring hourly solar radiation as of 1967.

sometimes inclined at a fixed slope, or normal;
5. If averaged, the period over which they are averaged
(e.g., monthly averages of daily radiation).

Examples of various kinds of radiation data are shown in
Tables 3.3.1, 3.3.2, and 3.3.3. Table 3.3.1 indicates daily
radiation on horizontal surface, averaged by months, for several
locations; this is the most common form in which data are available, for example, from Löf et al. (1966a). Table 3.3.2 shows
hourly average radiation, by months, for Madison. Table 3.3.3
is a sample of integrated hourly values of radiation and temperature.
    Averaged solar radiation data are also available from maps
that indicate general trends and averages. For example, world
maps are shown in Figures 3.3.1 to 3.3.4 [Löf et al. (1966a or
b)].* These are useful in predicting areas of potential solar
energy applications. In some geographical areas where climate
does not change abruptly with distance (i.e., away from major
influences such as mountains or large industrial cities), maps
can be used as a source of average radiation if data are not
available. However, large-scale maps must be used with care,
because they do not show local physical or climatological conditions which may greatly affect local solar energy availability.

3.4 ESTIMATION OF AVERAGE SOLAR RADIATION

In addition to solar radiation data, there are other meteorological measurements which are related to solar energy, and which
in the absence of radiation data can be used to estimate radiation. Data on hours of sunshine, or percent of possible sunshine hours, are widely available from many hundreds of stations
in many countries. Examples are shown in Table 3.4.1. Cloud
cover data (i.e., cloudiness) are also available. The use of
these measurements in estimating average radiation is discussed
below.
    Radiation data are the best source of information. Lacking
these, it is possible to use empirical relationships to estimate
radiation from hours of sunshine or percent possible sunshine,
or cloudiness. A third alternative is estimation for a particular location by use of data from other locations of similar latitude, topography, and climate.
    The original Angstrom-type regression equation related mean

*Figures 3.3.1 to 3.3.4 are reproduced from deJong (1973), who
redrew maps originally published by Löf, et. al. deJong has
compiled maps and radiation data from many sources.

TABLE 3.3.1 Typical Solar Radiation Data, cal/cm² day, on Horizontal Surface, Averaged by Months.

Station	Latitude	Alt., m	Annual	Radiation											
				Jan	Feb	Mar	Apr	May	Jun	Jul	Aug	Sep	Oct	Nov	Dec
Albuquerque	35°N	1620	519	308	390	512	630	688	739	690	636	566	447	338	281
Atlanta	33°N	308	394	228	284	377	484	535	554	538	502	412	350	265	201
Blue Hill	42°N	194	322	156	215	304	379	471	517	500	434	352	249	158	129
Charleston	32°N	18	404	253	303	392	515	550	560	523	496	408	343	285	216
Columbia	38°N	248	381	180	250	344	433	531	571	574	526	448	324	222	169
Fresno	36°N	110	446	186	296	438	545	637	697	668	606	503	375	241	160
Gainesville	29°N	59	431	278	367	445	539	586	544	520	508	444	368	318	254
Lander	42°N	1699	441	230	321	451	553	585	675	648	580	464	356	237	197
Lincoln	40°N	360	368	190	255	347	424	496	545	537	508	412	352	207	172
Madison	43°N	271	339	155	220	324	388	476	555	553	471	384	265	150	132
San Antonio	29°N	249	439	277	347	419	450	539	604	625	582	490	395	291	252
State College	40°N	375	335	139	202	297	373	467	544	528	454	361	275	155	120
Pretoria	26°S	1418	475	610	520	490	410	360	340	360	430	500	530	570	580
Canberra	34°S	177	424	619	546	443	344	264	205	228	303	426	505	594	637
Tokyo	36°N	s.l.	261	190	231	274	312	343	303	336	338	254	202	185	169
Stockholm	59°N	s.l.	241	29	78	201	308	467	517	500	392	243	112	32	18

TABLE 3.3.2 Hourly Average Horizontal Radiation by Months for Madison, Wisconsin.

	Radiation, langleys in the hour ending at:													
Month	6	7	8	9	10	11	12	1	2	3	4	5	6	7
January	--	--	1	7	15	21	25	25	23	16	9	2	--	--
February	--	--	4	13	21	28	32	34	31	24	15	6	1	--
March	--	2	12	23	33	40	43	43	39	32	21	11	3	--
April	1	8	17	27	36	43	47	44	41	33	26	17	2	--
May	5	15	27	39	49	55	58	58	54	45	35	25	14	5
June	7	17	29	42	65	59	63	61	58	51	41	31	19	8
July	7	18	31	43	54	61	64	63	58	52	43	31	18	7
August	2	12	25	37	48	56	58	57	51	44	35	23	12	3
September	--	5	18	31	43	51	55	54	49	42	29	17	6	--
October	--	1	8	20	30	39	42	42	38	30	19	8	1	--
November	--	--	2	10	19	25	29	29	25	18	9	2	--	--
December	--	--	1	6	13	18	22	22	19	12	6	1	--	--

TABLE 3.3.3 Hourly Radiation, Air Temperature and Wind Speed Data for a January Week, Boulder, Colorado.*

Day	Hour	H	°C	V	Day	Hour	H	°C	V
8	1	0.	-1.7	3.1	8	13	1105.	2.8	8.0
8	2	0.	-3.3	3.1	8	14	1252.	3.8	9.8
8	3	0.	-2.8	3.1	8	15	641.	3.3	9.8
8	4	0.	-2.2	3.1	8	16	167.	2.2	7.2
8	5	0.	-2.8	4.0	8	17	46.	.6	7.6
8	6	0.	-2.8	3.6	8	18	0.	-.6	7.2
8	7	0.	-2.2	3.6	8	19	0.	-1.1	8.0
8	8	17.	-2.2	4.0	8	20	0.	-1.7	5.8
8	9	134.	-1.1	1.8	8	21	0.	-1.7	5.8
8	10	331.	1.1	3.6	8	22	0.	-2.2	7.2
8	11	636.	2.2	1.3	8	23	0.	-2.2	6.3
8	12	758.	2.8	2.2	8	24	0.	-2.2	5.8

31

(continued)

TABLE 3.3.3 (Continued)

Day	Hour	H	°C	V	Day	Hour	H	°C	V
9	1	0.	-2.8	7.2	9	13	1185.	-2.2	2.2
9	2	0.	-3.3	7.2	9	14	1009.	0.0	0.0
9	3	0.	-3.3	6.3	9	15	796.	-.6	1.3
9	4	0.	-3.3	5.8	9	16	389.	-.6	1.3
9	5	0.	-3.9	4.0	9	17	134.	-2.2	4.0
9	6	0.	-3.9	4.5	9	18	0.	-2.8	4.0
9	7	0.	-3.9	1.8	9	19	0.	-3.3	4.5
9	8	4.	-3.9	2.2	9	20	0.	-5.6	5.8
9	9	71.	-3.9	2.2	9	21	0.	-6.7	5.4
9	10	155.	-3.3	4.0	9	22	0.	-7.8	5.8
9	11	343.	-2.8	4.0	9	23	0.	-8.3	4.5
9	12	402.	-2.2	4.0	9	24	0.	-8.3	6.3
10	1	0.	-9.4	5.8	10	13	1872.	2.2	7.6
10	2	0.	-10.0	6.3	10	14	1733.	4.4	6.7
10	3	0.	-8.9	5.8	10	15	1352.	6.1	6.3
10	4	0.	-10.6	6.3	10	16	775.	6.7	4.0
10	5	0.	-8.3	4.9	10	17	205.	6.1	2.2
10	6	0.	-8.3	7.2	10	18	4.	3.3	4.5
10	7	0.	-10.0	5.8	10	19	0.	.6	4.0
10	8	33.	-8.9	5.8	10	20	0.	.6	3.1
10	9	419.	-7.2	6.7	10	21	0.	0.0	2.7
10	10	1047.	-5.0	9.4	10	22	0.	.6	2.2
10	11	1570.	-2.2	8.5	10	23	0.	1.7	3.6
10	12	1805.	-1.1	8.0	10	24	0.	.6	2.7
11	1	0.	-1.7	8.9	11	13	138.	-5.0	6.7
11	2	0.	-2.2	4.9	11	14	96.	-3.9	6.7
11	3	0.	-2.2	4.5	11	15	84.	-4.4	7.6
11	4	0.	-2.8	5.8	11	16	42.	-3.9	6.3
11	5	0.	-4.4	5.4	11	17	4.	-5.0	6.3
11	6	0.	-5.0	4.5	11	18	0.	-5.6	4.5
11	7	0.	-5.6	3.6	11	19	0.	-6.7	4.5
11	8	4.	-6.1	5.8	11	20	0.	-7.8	3.1
11	9	42.	-5.6	5.4	11	21	0.	-9.4	2.7
11	10	92.	-5.6	5.4	11	22	0.	-8.9	3.6
11	11	138.	-5.6	9.4	11	23	0.	-9.4	4.0
11	12	163.	-5.6	8.0	11	24	0.	-11.1	3.1

(continued)

TABLE 3.3.3 (Continued)

Day	Hour	H	°C	V	Day	Hour	H	°C	V
12	1	0.	-11.7	4.0	12	13	389.	-2.2	5.8
12	2	0.	-12.8	3.1	12	14	477.	-.6	4.0
12	3	0.	-15.6	7.2	12	15	532.	2.8	2.2
12	4	0.	-16.7	6.7	12	16	461.	-.6	2.2
12	5	0.	-16.7	6.3	12	17	33.	-1.7	3.1
12	6	0.	-16.1	6.3	12	18	0.	-4.4	1.3
12	7	0.	-17.2	3.6	12	19	0.	-7.8	2.7
12	8	17.	-17.8	2.7	12	20	0.	-7.8	4.0
12	9	71.	-13.3	8.0	12	21	0.	-8.9	4.9
12	10	180.	-11.1	8.9	12	22	0.	-10.6	4.9
12	11	247.	-7.8	8.5	12	23	0.	-12.8	4.9
12	12	331.	-5.6	7.6	12	24	0.	-11.7	5.4
13	1	0.	-10.6	4.0	13	13	1926.	5.6	5.4
13	2	0.	-10.6	5.4	13	14	1750.	7.2	4.5
13	3	0.	-10.0	4.5	13	15	1340.	8.3	4.9
13	4	0.	-11.1	3.1	13	16	703.	8.9	4.5
13		0.	-10.6	3.6	13	17	59.	6.7	5.4
13	6	0.	-9.4	3.1	13	18	0.	4.4	3.6
13	7	0.	-7.2	3.6	13	19	0.	1.1	3.6
13	8	17.	-10.6	4.0	13	20	0.	0.0	3.1
13	9	314.	-8.3	5.8	13	21	0.	-2.2	6.7
13	10	724.	-1.7	6.7	13	22	0.	2.8	7.2
13	11	1809.	1.7	5.4	13	23	0.	1.7	8.0
13	12	2299.	3.3	6.3	13	24	0.	1.7	5.8
14	1	0.	-.6	7.2	14	13	1968.	6.7	1.8
14	2	0.	-1.1	7.6	14	14	1733.	6.7	2.7
14	3	0.	-.6	6.3	14	15	1331.	7.2	3.1
14	4	0.	-3.9	2.7	14	16	837.	6.7	3.1
14	5	0.	-1.7	4.9	14	17	96.	7.2	2.7
14	6	0.	-2.8	5.8	14	18	4.	3.3	2.7
14	7	0.	-2.8	4.0	14	19	0.	0.0	3.6
14	8	38.	-5.0	3.1	14	20	0.	3.9	5.4
14	9	452.	-5.0	4.9	14	21	0.	-3.9	3.6
14	10	1110.	-1.7	4.5	14	22	0.	-3.9	5.8
14	11	1608.	2.8	3.1	14	23	0.	-6.1	5.4
14	12	1884.	3.8	3.6	14	24	0.	-6.7	6.3

*Solar radiation, H, is kJ/m^2 for the hour ending at the indicated time; wind speed, V, is m/s. (See Figure 10.2.2.)

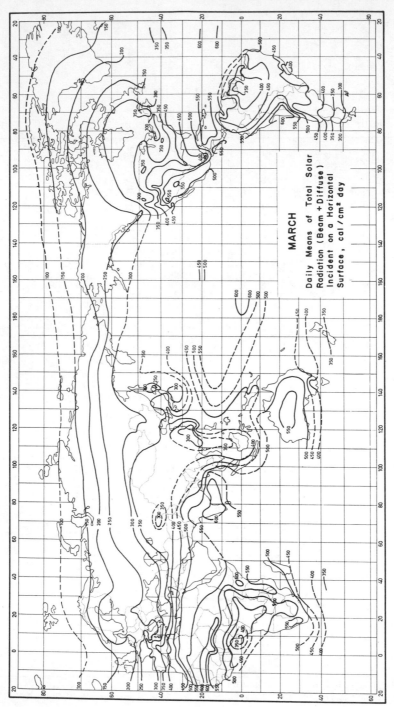

MARCH

Daily Means of Total Solar
Radiation (Beam + Diffuse)
Incident on a Horizontal
Surface, cal/cm² day

Figure 3.3.1 Daily radiation for March.

34

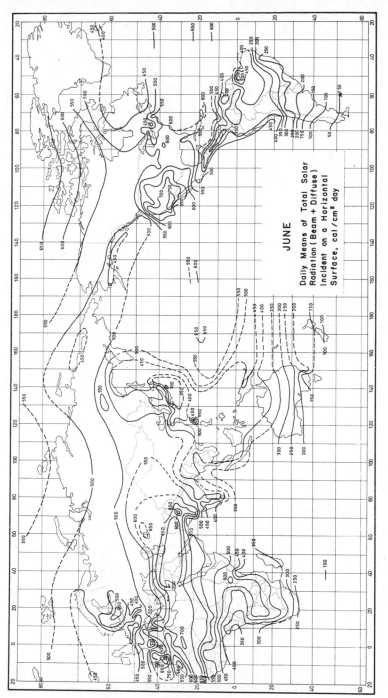

Figure 3.3.2 Daily radiation for June.

35

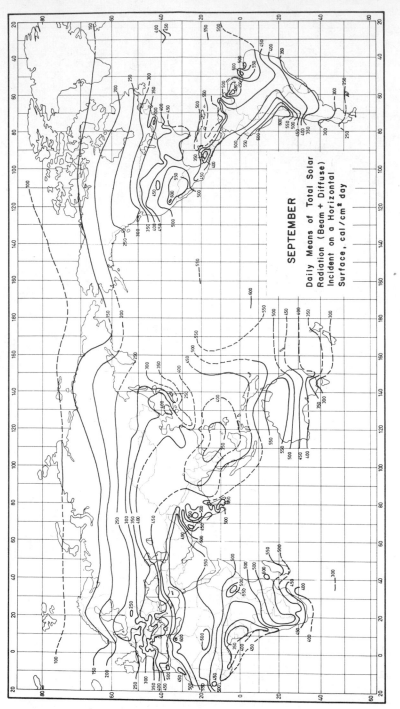

SEPTEMBER

Daily Means of Total Solar
Radiation (Beam + Diffuse)
Incident on a Horizontal
Surface, cal/cm² day

Figure 3.3.3 Daily radiation for September.

36

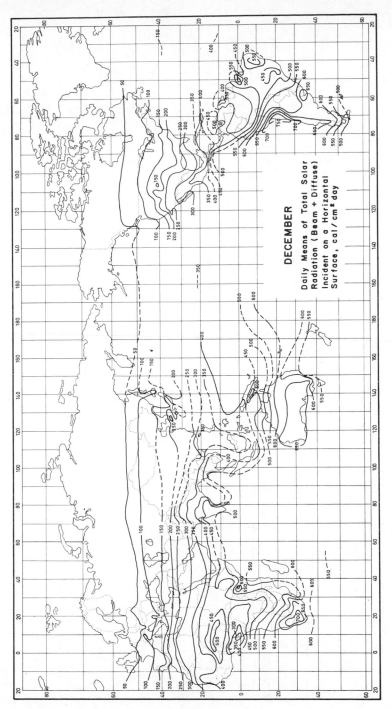

DECEMBER

Daily Means of Total Solar
Radiation (Beam + Diffuse)
Incident on a Horizontal
Surface, cal/cm²·day

Figure 3.3.4 Daily radiation for December.

TABLE 3.4.1 Examples of Monthly Averages of Hours Per Day of Sunshine.

Station	Latitude	Alt., m	Annual	Jan.	Feb.	Mar.	Apr.	May	June	July	Aug.	Sept.	Oct.	Nov.	Dec.
Hong Kong	22°N	s.l.	5.3	4.7	3.5	3.1	3.8	5.0	5.3	6.7	6.4	6.6	6.8	6.4	5.6
Paris	48°N	50	5.1	2.1	2.8	4.9	7.4	7.1	7.6	8.0	6.8	5.6	4.5	2.3	1.6
Bombay	19°N	s.l.	7.4	9.0	9.3	9.0	9.1	9.3	5.0	3.1	2.5	5.4	7.7	9.7	9.6
Sokoto (Nigeria)	13°N	107	8.8	9.9	9.6	8.8	8.9	8.4	9.5	7.0	6.0	7.9	9.6	10.0	9.8
Perth (Australia)	32°S	20	7.8	10.4	9.8	8.8	7.5	5.7	4.8	5.4	6.0	7.2	8.1	9.6	10.4
Madison	43°N	63	7.3	4.5	5.7	6.9	7.5	9.1	10.1	9.8	10.0	8.6	7.2	4.2	3.9

radiation to clear day radiation (at the location in question) and mean fraction of possible sunshine hours:

$$H_{av} = H_0' \left(a' + b' \frac{n}{N}\right) \tag{3.4.1}$$

where H_{av} = average horizontal radiation for the period in
              question (e.g., month);
         H_0' = clear day horizontal radiation for the same
              period;
         n = average daily hours of bright sunshine for same
              period;
         N = maximum daily hours of bright sunshine for same
              period;
      a', b' = constants.

Several authors have used sunshine and radiation data for the same stations and times to statistically derive values of a' and b'. Fritz (1951) gives $a' = 0.35$ and $b' = 0.61$. Values of H_0' for use in Eq. (3.4.1) can be obtained from charts such as Figure 3.4.1. It is also necessary to know the day length; this can be calculated from Eq. (2.5.5) or it can conveniently be obtained from Figure 3.4.2, a nomogram developed by Whillier (1965) showing day length as a function of declination and latitude.

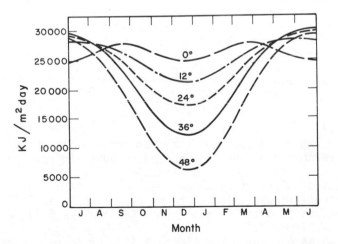

Figure 3.4.1 Clear day solar radiation on a horizontal plane for various latitudes.

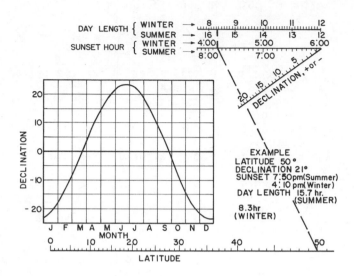

Figure 3.4.2 Nomogram to determine hour of sunset and day length [adapted from Whillier (1965)].

A basic difficulty with Eq. (3.4.1) lies in the ambiguity of the terms H_0 and n/N. The latter is an instrumental problem (records from sunshine recorders are open to interpretation), while the former stems from lack of definition of what constitutes a clear day. More recently Page (1964) and others have modified the method to base it on extraterrestrial radiation on a horizontal surface.

$$H_{av} = H_0 \left(a + b \, \frac{n}{N} \right) \qquad (3.4.2)$$

where H_0 = radiation outside of the atmosphere for the same location, averaged over the time period in question and a,b = modified constants depending on location.

$a, b \rightarrow$ Löf et al. (1966a) developed sets of constant a and b for various locations and climate types; these are given in Table 3.4.2. Values of N can be determined from Eq. (2.5.5), or obtained from Figure 3.4.2.

H_0 can be obtained from Figure 3.4.3, or it can be calculated by

$$H_0 = \frac{24}{\pi} I_{sc} \left(\left[1 + 0.033 \cos \left(\frac{360n}{365} \right) \right] \left[\cos \phi \cos \delta \sin \omega_s \right. \right.$$

$$\left. \left. + \frac{2\pi\omega_s}{360} \sin \phi \sin \delta \right] \right) \tag{3.4.3}$$

where I_{sc} = solar constant, per hour, n = day of the year, and ω_s = sunrise hour angle. The declination can be obtained from Eq. (2.5.1) and the sunrise hour angle from Eq. (2.5.5).

Example 3.4.1 Estimate the monthly averages of total solar radiation on a horizontal surface for Madison, Wisconsin, latitude 43°N, based on the average duration of sunshine hours data of Table 3.4.1.

The estimates are based on Eq. 3.4.2 using constants a = 0.30 and b = 0.34 from Table 3.4.2. Values of H_0 are obtained from Figure 3.4.3 and day lengths N from Figure 3.4.2, each for the midpoints of the month. The desired estimates are obtained in the following table, which shows H_{av} in kJ/m² day.

Month	H_0, 10^3 kJ/m² day	δ	N hr	n/N	H_{av} = $(0.30 + 0.34n/N)H_0$ 10^3 kJ/m² day
Jan	13.0	-21.3	9.2	0.489	6.1
Feb	18.4	-13.3	10.3	0.553	9.0
Mar	25.4	-2.8	11.7	0.590	12.7
Apr	33.4	9.4	13.2	0.568	16.5
May	39.0	18.8	14.5	0.628	20.0
June	41.4	23.3	15.2	0.665	21.8
July	40.3	21.5	14.9	0.658	21.1
Aug	35.7	13.8	13.8	0.725	19.5
Sept	28.5	2.2	12.3	0.699	15.3
Oct	20.7	-9.6	10.8	0.667	10.9
Nov	14.3	-19.2	9.5	0.442	6.4
Dec	11.6	-23.3	8.8	0.443	5.2

Some data are also available on tenths of cloudiness, C. It is also possible to derive an empirical relationship between

TABLE 3.4.2 Climatic Constants for Use in Eq. (3.4.2).

Location	Climate*	Veg.[†]	Sunshine Hours in Percentage of Possible		a	b
			Range	Avg.		
Albuquerque, N. M.	BS-BW	E	68-85	78	.41	.37
Atlanta, Ga.	Cf	M	45-71	59	.38	.26
Blue Hill, Mass.	Df	D	42-60	52	.22	.50
Brownsville, Tex.	BS	GDsp	47-80	62	.35	.31
Buenos Aires, Arg.	Cf	G	47-68	59	.26	.50
Charleston, S. C.	Cf	E	60-75	67	.48	.09
Darien, Manchuria	Dw	D	55-81	67	.36	.23
El Paso, Tex.	BW	Dsi	78-88	84	.54	.20
Ely, Nevada	Bw	Bzi	61-89	77	.54	.18
Hamburg, Germany	Cf	D	11-49	36	.22	.57
Honolulu, Hawaii	Af	G	57-77	65	.14	.73
Madison, Wisconsin	Df	M	40-72	58	.30	.34
Malange, Angola	Aw-BS	GD	41-84	58	.34	.34
Miami, Fla.	Aw	E-GD	56-71	65	.42	.22
Nice, France	Cs	SE	49-76	61	.17	.63
Poona, India	Am	S	25-49	37	.30	.51
(Monsoon Dry)			65-89	81	.41	.34
Stanleyville, Congo	Af	B	34-56	48	.28	.39
Tamanrasset, Algeria	BW	Dsp	76-88	83	.30	.43

*Climatic classification based on Trewartha's climate map (1954, 1961), where climate types are:

Af -- Tropical forest climate, constantly moist; rainfall all through the year;
Am -- Tropical forest climate, monsoon rain; short dry season, but total rainfall sufficient to support rain forest;
Aw -- Tropical forest climate, dry season in winter;
BS -- Steppe or semiarid climate;
BW -- Desert or arid climate;
Cf -- Mesothermal forest climate; constantly moist; rainfall all through the year;
Cs -- Mesothermal forest climate; dry season in winter;
Df -- Microthermal snow forest climate; constantly moist; rainfall all through the year;
Dw -- Microthermal snow forest climate; dry season in winter.

(continued)

TABLE 3.4.2 (Continued)

[†]Vegetation classification based on Küchler's map, where vegetation types are:

B   -- Broadleaf evergreen trees;
Bzi   -- Broadleaf evergreen, shrubform, minimum height 3 feet, growth singly or in groups or patches;
D   -- Broadleaf deciduous trees;
Dsi   -- Broadleaf deciduous, shrubform, minimum height 3 feet, plants sufficiently far apart that they frequently do not touch;
Dsp   -- Broadleaf deciduous, shrubform, minimum height 3 feet, growth singly or in groups or patches;
E   -- Needleleaf evergreen trees;
G   -- Grass and other herbaceous plants;
GD   -- Grass and other herbaceous plants; broadleaf deciduous trees;
GDsp   -- Grass and other herbaceous plants; broadleaf deciduous, shrubforms, minimum height 3 feet, growth singly or in groups or patches;
M   -- Mixed: broadleaf deciduous and needleleaf evergreen trees;
S   -- Semideciduous: broadleaf evergreen and broadleaf deciduous trees;
SE   -- Semideciduous: broadleaf evergreen and broadleaf deciduous trees; needleleaf evergreen trees.

monthly averages of C and average radiation of a form such as

$$\frac{H_{av}}{H_0} = a'' - b''C \qquad (3.4.4)$$

This procedure is less satisfactory than that based on Eq. (3.4.2). Large scatter is to be expected for individual days, and it is suggested that this be used only as a last resort. See Norris (1968) for a further discussion.

3.5 *ESTIMATION OF HOURLY RADIATION FROM DAILY DATA*

For purposes of design and evaluation of solar processes, it is often desired to make hour-by-hour (or other short time base) performance calculations for the system. Radiation data are

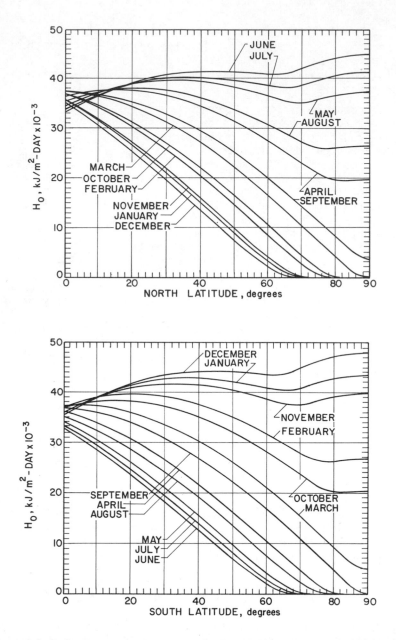

Figure 3.4.3 Extraterrestrial daily insolation on a horizontal surface, for mid points of months.

44

available on an hourly basis for some U.S. stations, but it is often necessary to start with daily data and estimate hourly values from the daily numbers. It must be recognized that this is not an exact process. For example, daily total radiation values in the middle range between clear day and completely cloudy day values can arise from various circumstances, such as intermittent heavy clouds, continuous light clouds, or heavy cloud cover for part of the day. There is no way to determine these circumstances from the daily totals. However, the methods presented here work best for clear days, because those are the days that produce most of the output of solar processes (particularly those processes that operate at temperatures significantly above ambient).

Statistical studies of the time distribution of total radiation on horizontal surfaces through the day, using average data from a number of stations, have led to generalized charts of the ratio of hourly total to daily total radiation as a function of day lengths and the hour in question. Figure 3.5.1 (solid curves) shows such a chart from Liu and Jordan (1967) based on Whillier (1956, 1965) and on Hottel and Whillier (1958). Day length can be calculated from Eq. (2.5.6) or it can be conveniently determined from Figure 3.4.2. Thus, from a knowledge of day length (a function of latitude θ and declination δ) and daily total radiation, the hourly radiation can be estimated.

Figure 3.5.1 is based on long-term averages and is intended for use in determining averages of hourly radiation. For lack of a useful alternative, it can be applied to individual days, with best results for clear days and increasingly uncertain results as daily total radiation decreases.

Example 3.5.1 The total radiation for Madison on August 23, 1956, was 31,400 kJ/m². The radiation received between 1 and 2 p. m. can be estimated as follows:

For that date, $\delta = 12°$. ϕ for Madison is 43°N. From Figure 3.4.2, sunset is at 6:45 p. m. and day length is 13.5 hr. Then from Figure 3.5.1, at day length of 13.5 hr and mean of 1½ hr from solar noon, the ratio of hourly total to daily total is 0.115. The estimated radiation in the hour from 1 to 2 p. m. is then 3600 kJ/m². (The measured value for that hour is 3470 kJ/m².)

It is also possible, on the basis of an analysis by Liu and Jordan (1960), to estimate the average hourly diffuse radiation from the average daily diffuse radiation. They have shown, for horizontal surfaces at a particular station, that the ratio

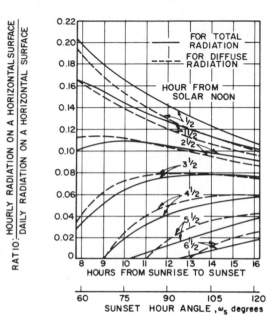

Figure 3.5.1   Relationships between daily radiation and hourly radiation on a horizontal surface.   From Liu and Jordan (1963). (Use curves for diffuse radiation in conjunction with Figure 3.5.3 for monthly averages.   If applied to individual clear days, use in conjunction with Figure 3.5.2.)

of daily diffuse radiation to the daily total radiation for individual days is nearly a unique function of a cloudiness index, that is, the ratio of daily total radiation to extraterrestrial daily insolation, H/H_0.   This is shown in Figure 3.5.2. Similarly, a plot of the ratios of monthly averages of daily diffuse radiation to monthly averages of daily total radiation as a function of average cloudiness, $\bar{H}/\bar{H}_0$, is shown in Figure 3.5.3.

Figure 3.5.1 also shows a set of curves (dashed) for diffuse radiation, indicating the hourly diffuse as a fraction of daily diffuse as a function of time and day length.   It is based on monthly average data for two stations.   These curves are based on averages and are not intended to be applied to individual days.   In conjunction with Figure 3.5.3, it can be

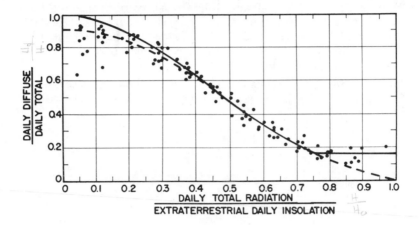

Figure 3.5.2 The ratio of the daily diffuse radiation to the daily total radiation as a function of cloudiness index. From Liu and Jordan (1960).

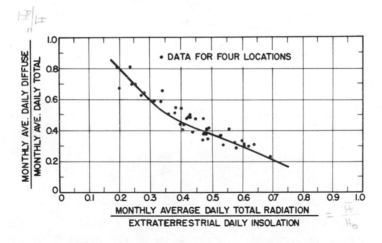

Figure 3.5.3 The ratio of monthly averages of daily diffuse to daily total radiation as a function of average monthly cloudiness index. From Liu and Jordan (1960).

47

used to estimate hourly averages of diffuse radiation if the average daily total radiation is known.

Example 3.5.2 The average June total radiation on a horizontal plane in Madison is 537 cal/cm^2 day (i.e., 22,500 kJ/m^2 day) (from Table 3.3.1). Estimate the average diffuse radiation and the average total radiation for the hours 10 to 11 and 1 to 2.

The extraterrestrial radiation H_0 for June for Madison is 41,400 kJ/m^2 day and the day length is 15.1 hr (from Example 3.4.1). Then

$$\frac{\bar{H}}{H_0} = \frac{537 \times 41.9}{41,400} = 0.54$$

and from Figure 3.5.3

$$\frac{H_d}{\bar{H}} = 0.33$$

so the average daily diffuse radiation is 0.33 × 22,500 = 7400 kJ/m^2 day. Entering Figure 3.5.1 for an average day length of 15.1 hr and for 1.5 hr from solar noon, using the dashed curve:

$$\text{Ratio } \frac{\text{average hourly diffuse radiation}}{\text{average daily diffuse radiation}} = 0.102$$

Thus the average diffuse for those hours is 0.102 × 7400 = 750 kJ/m^2.

Again from Figure 3.5.1, from the solid curve for 1.5 hr from solar noon, for an average day length of 15.1 hr:

$$\text{Ratio } \frac{\text{average hourly total radiation}}{\text{average daily total radiation}} = 0.108$$

and average hourly total = 0.108 × 22,500 = 2400 kJ/m^2.

3.6 *RATIO OF BEAM RADIATION ON TILTED SURFACE TO THAT ON HORIZONTAL SURFACE*

For purposes of solar process design, it is often necessary to convert data for hourly radiation (measured or estimated by methods of section 3.5) on a horizontal surface to radiation on a tilted surface. This can be done exactly for the beam component, and for diffuse radiation coming from the part of the sky around the sun. The problem of estimating a conversion for the diffuse component will be discussed in section 3.7. From Figure 3.6.1, it follows* that $H = H_n \cos \theta_z$ and $H_T = H_n \cos \theta_T$. The ratio of radiation on the tilted surface, H_T, to that on the horizontal surface, H, is given in terms of the angles θ_z and θ_T and radiation normal to the beam, H_n, by

$$R_b = \frac{H_T}{H} = \frac{H_n \cos \theta_T}{H_n \cos \theta_z} = \frac{\cos \theta_T}{\cos \theta_z} \qquad (3.6.1)$$

where $\cos \theta_T$ and $\cos \theta_z$ are both found from Eq. (2.5.2).

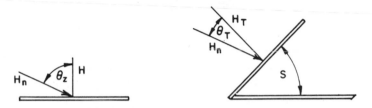

Figure 3.6.1   Radiation on horizontal and tilted surfaces.

Here we outline a convenient method from Hottel and Woertz (1942) for surfaces tilted toward the equator. Equations (2.5.3) and (2.5.4) can be used to indicate the angle of incidence of beam radiation on both the horizontal and tilted surfaces (if the tilted surface has its inclination north or south):

$$R_b = \frac{\cos \theta_T}{\cos \theta_z} = \frac{\cos(\phi - s)\cos \delta \cos \omega + \sin(\phi - s)\sin \delta}{\cos \phi \cos \delta \cos \omega + \sin \phi \sin \delta}$$
$$(3.6.2)$$

Equation (3.6.2) can be used directly, or convenient graphical solutions of Hottel and Woertz can be used. Figures 3.6.2 show plots of $\cos \theta_T$ as a function of declination and

*Note that the symbol H used here is meant to imply that portion of the total solar radiation for which the angular correction is made, strictly speaking, the beam component only.

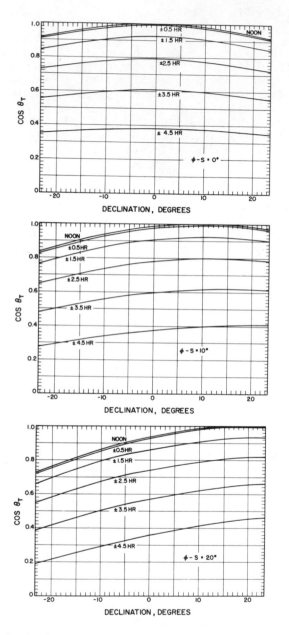

Figure 3.6.2 Cosine of angle of incidence of beam radiation on surfaces tilted toward the equator for various values of $(\phi - \varepsilon)$. From Hottel and Woertz (1942).

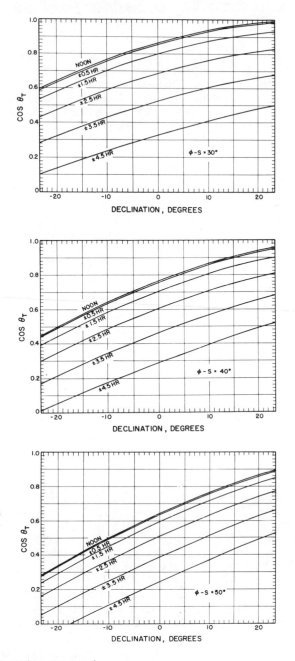

Figure 3.6.2 (Continued)

51

hour, for values of $(\phi - s)$ from $0°$ to $50°$. Values of $\cos \theta_z$
are obtained from these plots where $(\phi - s)$ equals the latitude
of the place in question. Values of $\cos \theta_T$ are obtained from
the plot with the appropriate $(\phi - s)$. The ratio R_b then follows. Values of $\cos \theta_T$ for negative value of $(\phi - s)$ are obtained from charts for the positive value, but with the sign
of the declination reversed. A plot of R_b for $\phi = 40°$ and
$s = 30°$ is given in Figure 3.6.3.

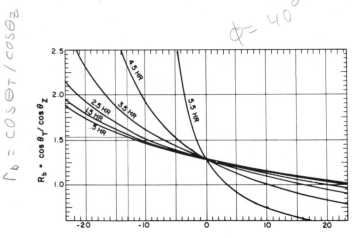

DECLINATION , DEGREES

Figure 3.6.3   Ratio R_b, $\cos \theta_T/\cos \theta_z$, for a surface with slope
$30°$ toward equator at latitude $40°$, at various times from noon.

Example 3.6.1 Suppose the latitude is $40°N$ and the tilt
is $30°$ toward the equator. Find the ratio, R_b for 9:30
on February 20. From Figure 3.4.2 the declination is $-12°$.
At 9:30, ω is equal to $37.5°$. Since $\phi - s$ is $10°$, $\cos \theta_T$
from Figure 3.6.2*b* is 0.73, since ϕ is $40°$, $\cos \theta_z$ from
Figure 3.6.2*e* is 0.46. The ratio, R_b is then 0.73/0.46
= 1.59, which is also given in Figure 3.6.3. Alternatively from Eq. (3.6.2),

$$R_b = \frac{\cos 10 \cos(-12) \cos 37.5 + \sin 10 \sin(-12)}{\cos 40 \cos(-12) \cos 37.5 + \sin 40 \sin(-12)} = 1.58$$

3.7 RATIO OF TOTAL RADIATION ON A TILTED SURFACE TO THAT ON A HORIZONTAL SURFACE

Flat-plate solar collectors absorb both beam and diffuse components of solar radiation. In order to use horizontal total radiation data, it is necessary to derive a value for R for total solar radiation. The angular correction is straightforward for the beam component. The correction for the diffuse component depends on the distribution of diffuse radiation over the sky, which generally is not known; this distribution depends particularly on clouds and also on the spatial distribution and amounts of other atmospheric components that determine scattering. Also, some solar radiation may be reflected from the ground to the surface. Two "limiting cases" have been assumed as a basis for angular correction of the diffuse radiation.

First, it can be assumed that most of the diffuse radiation comes from an apparent origin near the sun, that is, the scattering of solar radiation is mostly forward scattering. This approximation will be best on clear days. The angular correction factor to be applied to the diffuse component is essentially the same as that for the beam component, the horizontal radiation is treated as though it is all beam radiation, and $R = R_b$. Example 3.6.1 shows a sample calculation of R_b which applies to the total radiation under this assumption.

Second, it can be assumed that the diffuse component is uniformly distributed over the sky. This may be a reasonable approximation when, for example, there is a uniform cloud cover or when conditions are very hazy. Then the diffuse radiation on a surface of other than horizontal orientation is dependent only on how much of the sky the surface sees. If it is further assumed that the properties of the ground or other surfaces seen by a tilted surface reflect solar radiation in such a way as to be a source of diffuse solar radiation equivalent to the sky, then the surface will receive the same diffuse radiation no matter what its orientation. Under this assumption, the correction factor to convert for diffuse radiation is always unity. The radiation on a tilted surface, under these conditions, is

$$H_T = H_b R_b + H_d \qquad (3.7.1)$$

where H_b and H_d are the beam and diffuse components of solar radiation on the horizontal surface, and R_b is the beam correction factor from eq. (3.6.2) or its equivalent. The effective ratio of solar energy on the tilted surface to that on the horizontal surface is then

$$R = \frac{H_T}{H} = \frac{H_b}{H} R_b + \frac{H_d}{H} \qquad (3.7.2)$$

Example 3.7.1 For the conditions of Example 3.6.1, if 0.8 of the total solar radiation is beam radiation and 0.2 is diffuse (i.e., skies are quite clear), estimate R, assuming the diffuse component is uniformly incident on the surface from sky and ground.

Solution. From Eq. (3.7.2), and using the R_b calculated from Example 3.6.1: $R = 0.8(1.58) + 0.2 = 1.46$.

An improvement on this model has been derived by Liu and Jordan (1963) by considering the radiation on the tilted surface to be made up of three components: the beam radiation, diffuse solar radiation, and solar radiation reflected from the ground which the tilted surface "sees." A surface tilted at slope s from the horizontal sees a portion of the sky dome given by $(1 + \cos s)/2$; if the diffuse solar radiation is uniformly distributed over the sky dome, this is also the conversion factor for diffuse radiation. The tilted surface also sees ground, or other surroundings, and if those surroundings have a diffuse reflectance of ρ for solar radiation, the reflected radiation from the surrounding on the surface from total solar radiation is $(H_b + H_d)(1 - \cos s)\rho/2$. Combining the three terms, the total solar radiation on the tilted surface at any time is

$$H_T = H_b R_b + H_d \frac{1 + \cos s}{2} + (H_b + H_d)\frac{(1 - \cos s)\rho}{2} \qquad (3.7.3)$$

and by definition of R:

$$R = \frac{H_b}{H} R_b + \frac{H_d}{H} \frac{1 + \cos s}{2} + \frac{(1 - \cos s)\rho}{2} \qquad (3.7.4)$$

Liu and Jordan suggest values of ground reflectance of 0.2 when there is no snow and 0.7 when there is a snow cover.

Example 3.7.2 What will be the R factor for the conditions of Example 3.6.1, if 0.8 of the total radiation is beam and 0.2 is diffuse, and if reflectance of the ground is taken into account? Calculate R for bare ground and for snow.

Solution. Again using the beam radiation correction factor of Example 3.6.1 and Eq. (3.7.4), for the condition of no snow,

$$R = 0.8(1.58) + 0.2 \frac{1 + \cos 30}{2} + \frac{(1 - \cos 30)0.2}{2} = 1.46$$

For the condition of snow,

$$R = 0.8(1.58) + 0.2 \frac{1 + \cos 30}{2} + \frac{(1 - \cos 30)0.7}{2} = 1.50$$

(Note that if s had been larger than 30°, the contribution of the third term becomes of increasing importance.)

None of these approximations is entirely satisfactory. The argument can be made that a solar collector will deliver the largest fraction of its total energy output during periods of high radiation, and that the first assumption is most nearly true under clear sky conditions. This argument is most valid when the collector is operating at temperatures substantially above ambient so that useful energy collection is not likely to occur during periods of extensive cloud cover.
Other approaches have been developed. For example, Heywood (1966) has made extensive simultaneous measurements in the London, England, area of total radiation on horizontal and inclined planes under cloudless sky conditions, and has developed an empirical method, based on those measurements, for estimating an angular correction factor for total radiation.

3.8 EFFECTS OF RECEIVING SURFACE ORIENTATION AND MOTION

Most solar radiation data are available on horizontal surfaces, but it is often desirable to know the effects of receiving-surface orientation. The methods of the previous section show how the effects of surface orientation can be estimated over short periods. For long periods, in locations where there are not marked seasonal variations in atmospheric conditions, estimates of the effects of orientation have been based on calculation of beam radiation. For example, Morse and Czarnecki (1968) calculated the relative annual insolation on surfaces facing the equator at various inclinations; their results are shown in Figure 3.8.1. They suggest that a slope of 0.9ϕ will result in the maximum annual incident beam radiation.

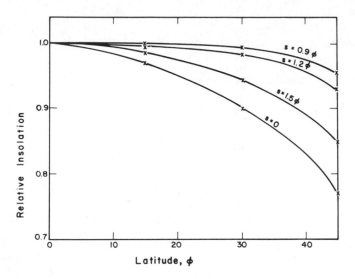

Figure 3.8.1 Relative annual insolation on surfaces of various inclinations s, for surfaces facing the equator ($\gamma = 0$). From Morse and Czarnecki (1958).

Others have reached similar conclusions, that is, for maximum annual collection, a receiver should be oriented toward the equator with a slope approximately equal to the latitude (i.e., $s = \phi$), for best winter radiation the slope should be about $\phi + 10°$, and for best summer radiation the slope should be about $\phi - 10°$. The curves of Figure 3.8.1 suggest that a few degrees variation in s from the optimum will have little effect on the total long time radiation.

Morse and Czarnecki also considered the question of effects of surface azimuth angle on total annual beam radiation, that is, the effects of orientation of the receiving surface east or west of due south (or north, in the southern hemisphere). They calculated the relative annual radiation on surfaces tilted at angles s of 0.9ϕ as a function of latitude ϕ and surface azimuth angle γ. These are shown in Figure 3.8.2. It can be seen that at fixed s the effect of γ increases with latitude, but that for $\gamma = 22.5°$ the relative annual insolation is within two percent of that for $\gamma = 0$ for latitudes up to $45°$. Not shown by annual radiation figures is the effect of azimuth angle γ on the diurnal distribution of radiation. Each $15°$ of surface azimuth angle will shift the daily distribution of available energy by about one hour, toward morning if γ is positive and toward afternoon if γ is negative.

The calculated annual totals of radiation indicated in Figures 3.8.1 and 3.8.2 are for beam radiation only, and when diffuse radiation is added, the effects of orientation of the surface may be significantly different. However, at the present time inadequate information is available to draw generalizations beyond those based on beam radiation.

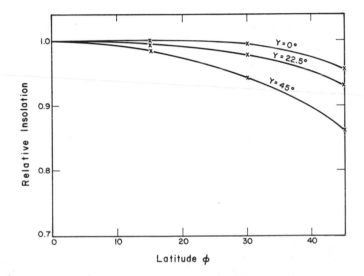

Figure 3.8.2 Relative annual insolation on surfaces at inclination γ = 0.9φ deviating from facing the equator by surface azimuth angle γ. From Morse and Czarnecki (1958).

It is also possible to estimate the annual beam radiation on surfaces moving in prescribed ways. This has been done by Eibling et al. (1953), for southwestern United States, using Moon's atmospheric transmission factors reduced to account for cloudiness. The results are shown in Table 3.8.1 for several modes of surface orientation.

TABLE 3.8.1 Calculated Effect of Surface Orientation on Annual Total and Beam Radiation, at 35° Latitude, Southwestern United States [From Eibling et al. (1953)].

Surface Orientation (Mode of Tracking)	Beam Radiation, 10^6 kJ/m^2
Fixed, horizontal	5.34
Fixed, tilted 35° south	6.19
Continuous adjustment about horizontal north-south axis	7.43
Continuous adjustment about axis parallel to the earth's axis	8.14
Continuous adjustment about two axes to maintain normal solar incidence	8.38

REFERENCES

deJong, B., Monograph published by Delft University Press, Netherlands, 1973. "Net Radiation Received by a Horizontal Surface at the Earth."

Drummond, A. J., Proceedings of the UN Conference on New Sources of Energy, *4*, 335 (1964).

Eibling, J. A., Thomas, R. E., and Landry, B. A., Report from Battelle Memorial Institute to U. S. Department of the Interior, December, 1953. "An Investigation of Multiple-Effect Evaporation of Saline Waters by Steam from Solar Radiation."

Fritz, S., <u>Compendium of Meteorology</u>, American Meteorological Society, 1951. "Solar Radiation Energy and its Modification by the Earth and its Atmosphere."

Heywood, H., Solar Energy, *10*, 51 (1966). "The Computation of Solar Radiation Intensities."

Hottel, H. C. and Whillier, A., <u>Transactions of the Conference on Use of Solar Energy</u>, *2*, 74, University of Arizona Press, 1958. "Evaluation of Flat Plate Solar Collector Performance."

Hottel, H. C. and Woertz, B. B., Trans. ASME *64*, 91 (1942). "Performance of Flat-Plate Solar-Heat Collectors."

Liu, B. Y. H. and Jordan, R. C., Solar Energy, *4*, No. 3, (1960). "The Interrelationship and Characteristic Distribution of Direct, Diffuse and Total Solar Radiation."

Liu, B. Y. H. and Jordan, R. C., Solar Energy, *7*, 53 (1963). "The Long-Term Average Performance of Flat-Plate Solar Energy Collectors."

Liu, B. Y. H. and Jordan, R. C., Low Temperature Engineering Applications of Solar Energy, New York, Am. Soc. of Heat., Ref. and Air Cond. Eng., 1967. "Availability of Solar Energy for Flat-Plate Solar Heat Collectors."

Löf, G. O. G., Duffie, J. A., and Smith, C. O., Engineering Experiment Station Report 21, University of Wisconsin, Madison (July 1966a). "World Distribution of Solar Radiation."

Löf, G. O. G., Duffie, J. A., and Smith, C. O., Solar Energy, *10*, 27 (1966b). "World Distribution of Solar Energy."

MacDonald, T. H., Mon. Weather Rev., *79*, No. 8, (1951). "Some Characteristics of the Eppley Pyrheliometer."

Morikofer, W., Transactions of the Conference on Use of Solar Energy, *1*, 60 and 63, University of Arizona Press, 1958. "On the Principles of Solar Radiation Measuring Instruments" and "Methods for the Open-Air Measurements of Caloric Radiation."

Morse, R. N. and Czarnecki, J. T., Report E. E. 6 of Engineering Section (now Mechanical Engineering Division), Commonwealth Scientific and Industrial Research Organization, Melbourne, Australia (1958). "Flat Plate Solar Absorbers: The Effect on Incident Radiation of Inclination and Orientation."

Norris, D. J., Solar Energy, *12*, 107 (1968). "Correlation of Solar Radiation with Clouds."

Page, J. K., Proceedings of the UN Conference on New Sources of Energy, *4*, 378 (1964). "The Estimation of Monthly Mean Values of Daily Total Short-Wave Radiation on Vertical and Inclined Surfaces from Sunshine Records for Latitudes 40°N-40°S."

Thekaekara, M. P., Solar Energy, *9*, 7 (1965). "The Solar Constant and Spectral Distribution of Solar Radiant Flux."

Trewartha, G. T., An Introduction to Climate, 3rd ed., New York, McGraw-Hill, 1954.

Trewartha, G. T., <u>The Earth's Problem Climates</u>, Madison, University of Wisconsin Press, 1961.

U. S. Weather Bureau, Climatological Data, National Summary.

U. S. Weather Bureau, National Data Center, Asheville, North Carolina.

Whillier, A., Arch. Met. Geoph. Biokl. B., *7*, 197, (1956). "The Determination of Hourly Values of Total Radiation from Daily Summations."

Whillier, A., Solar Energy, *9*, 165 (1965). "Solar Radiation Graphs."

Yellott, J. I., <u>Low Temperature Application of Solar Energy</u>, New York, Am. Soc. of Heat., Ref. and Air Cond. Eng., 1967. "The Measurement of Solar Radiation."

4. SELECTED TOPICS IN HEAT TRANSFER

This chapter is not intended to be a substitute for a textbook
on heat transfer but is intended to review those aspects of
heat transfer that are important in the design and/or analysis
of solar energy systems. The chapter will begin with a review
of radiation heat transfer, which is often given cursory treatment in standard heat transfer courses, and the following sections cover convection heat transfer between parallel plates.
The final sections review some basic correlations for internal
and external flow.
   The role of convection and conduction heat transfer in the
performance of solar systems is obvious. Radiation heat transfer plays a role in bringing energy to the earth, but not so
obvious is the significant role radiation heat transfer plays
in the operation of solar collectors. In the usual engineering
practice, radiation heat transfer is often negligible. In a solar collector, the energy flux is often two orders of magnitude
smaller than in conventional heat transfer equipment, and thermal radiation becomes a significant mode of heat transfer.
   In this review of radiation heat transfer, it will be necessary to utilize equations from quantum mechanics and electromagnetic theory. However, it is not necessary to be familar
with these topics in order to use the results in calculating
energy exchange by radiation heat transfer.

4.1 THE ELECTROMAGNETIC SPECTRUM

Radiation, as used in this book, is electromagnetic energy that
is propagated through space at the speed of light. For most
solar energy applications, only thermal radiation is important.
Thermal radiation is emitted by bodies by virtue of their temperature; the atoms, molecules, or electrons are raised to ex-
cited states, return spontaneously to lower energy states, and
in doing so, emit energy in the form of electromagnetic radiation. Because the emission results from changes in electronic,
rotational, and vibrational states of atoms and molecules, the
emitted radiation is usually distributed over a range of wavelengths.

61

The spectrum of electromagnetic radiation is divided into wavelength bands. These bands and the wavelengths representing their approximate limits, are shown in Figure 4.1.1. It should be noted that wavelength limits associated with the various names and the mechanism producing the radiation are not sharply defined. There is no basic distinction between these ranges of radiation other than the wavelength λ; they all travel with the speed of light C and have a frequency ν such that

$$C = \frac{C_0}{n} = \lambda\nu \qquad\qquad (4.1.1)$$

where C_0 is the speed of light in a vacuum and n is the index of refraction.

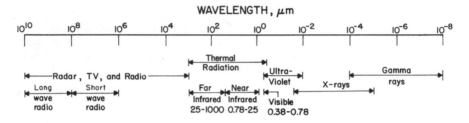

Figure 4.1.1 Spectrum of electromagnetic radiation.

The wavelengths of importance in solar energy and its applications are in the ultraviolet to near-infrared range, that is, from 0.2 to about 25 µm. This includes the visible spectrum, light being a particular portion of the electromagnetic spectrum to which the human eye responds. Solar radiation outside the atmosphere has most of its energy in the range of 0.2 to 4 µm, while solar energy received at the ground is substantially in the range of 0.29 to 3.0 µm.

4.2 PHOTON RADIATION

For some purposes in solar energy applications, the classical electromatic wave view of radiation does not explain the observed phenomena. In this connection, it is necessary to consider the energy of a particle or photon which can be thought of as an "energy unit" with zero mass and zero charge. The energy of the photon is given by

$$E = h\nu \qquad\qquad (4.2.1)$$

where h is Planck's constant (6.6256×10^{-34} J sec). It follows that as the frequency increases, that is, as the wavelength λ decreases, the photon energy increases. This fact is particularly significant where a minimum photon energy is needed to cause a required change. (For example, the creation of a hole-electron pair in a photovoltaic device.) There is thus an upper limit of wavelengths of radiation that can cause the change.

4.3 THE BLACKBODY, A PERFECT ABSORBER AND EMITTER OF RADIATION

By definition, a blackbody is a perfect absorber of radiation. No matter what wavelengths or directions describe the radiation incident on a blackbody, all incident radiation will be absorbed. A blackbody is an ideal concept since all real substances will reflect some radiation and/or permit some radiation to pass through.

Even though a true blackbody does not exist in nature, some materials approach a blackbody. For example, a large thickness of carbon black can absorb about 99% of all incident thermal radiation. This absence of reflected radiation is the reason for the name given to a blackbody. The eye would perceive a blackbody as being black. However, the eye is not a good indicator of the ability of a material to absorb radiation, since the eye is only sensitive to a small portion of the wavelength range of thermal radiation. White paint is a good reflector of visible radiation but is a good absorber of infrared radiation.

A blackbody is also a perfect emitter of thermal radiation. In fact, the definition of a blackbody could have been made in terms of a body that emits the maximum possible radiation. A simple thought experiment can be used to show that if a body is a perfect emitter of radiation, then it must also be a perfect absorber of radiation. Suppose a small blackbody and a small real body are placed in a large evacuated enclosure made from a blackbody material. If the enclosure is isolated from the surroundings, then the blackbody, the real body, and the enclosure will in time come to the same equilibrium temperature. Now, the blackbody must, by definition, absorb all the radiation incident on it, and in order to maintain a constant temperature, the blackbody must also emit an equal amount of energy. The real body in the enclosure must absorb less radiation than the blackbody and will consequently emit less radiation than the blackbody. Thus a blackbody both absorbs and emits the maximum amount of radiation.

4.4 PLANCK'S LAW AND WIEN'S DISPLACEMENT LAW

Radiation in the region of the electromagnetic spectrum from
about 0.2 to about 100 μm is called thermal radiation and is
emitted by all substances by virtue of their temperature. The
wavelength distribution of this radiation for a blackbody is
given by Planck's law* (Richtmyer and Kennard, 1947):

$$e_{b\lambda} = \frac{2\pi h C_0^2}{\lambda^5 \left(e^{hC_0/\lambda kT} - 1\right)} \tag{4.4.1}$$

where h is Planck's constant and k is Boltzmann's constant. The
groups $2\pi h C_0^2$ and hC_0/k are often called the first and second radiation constants and given the symbols, C_1 and C_2, respectively.
For calculations, $C_1 = 3.7405 \times 10^{-16}$ W m^2 and $C_2 = 0.0143879$ m °K.
   It is also of interest to know the wavelength corresponding
to the maximum intensity of blackbody radiation. By differenti-
ating Planck's distribution and equating to zero, the wavelength
corresponding to the maximum of the distribution can be derived.
This leads to Wien's displacement law, which can be written

$$\lambda_{max}T = 2897.8 \text{ } \mu\text{m °K} \tag{4.4.2}$$

   Planck's law and Wien's displacement law are illustrated
in Figure 4.4.1, which shows spectral radiation distribution
for blackbody radiation from sources at 6000, 1000, and 400 °K.
The shape of the distribution and the displacement of the wavelength of maximum intensity is clearly shown. Note that 6000 °K
represents an approximation of the surface temperature of the
sun, and that the distribution shown for that temperature is an
approximation of the distribution of solar radiation outside the
earth's atmosphere. The other two temperatures are representative of those encountered in low- and high-temperature solar-
heated surfaces.
   The same information as shown on Figure 4.4.1 has been replotted on a linear scale on Figure 4.4.2. The ordinate on this
figure, which ranges from zero to one, is the ratio of the spectral emissive power to the maximum value at the same temperature. This clearly shows the wavelength division between a
6000 °K source and lower temperature sources at 1000 °K and 400 °K.

*The symbol e represents energy per unit area per unit time per
unit wavelength interval at wavelength λ. The subscript b represents blackbody.

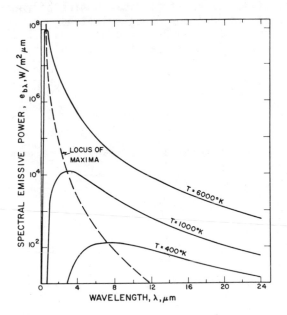

Figure 4.4.1   Spectral distribution of blackbody radiation.

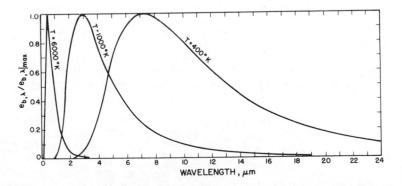

Figure 4.4.2   Normalized spectral distribution of blackbody radiation.

4.5 STEFAN-BOLTZMANN FORMULA

Planck's law gives the spectral distribution of radiation from a blackbody, but in engineering calculations, the total energy is often of more interest. By integrating Planck's law over all wavelengths, the total energy emitted by a blackbody is found to be

$$e_b = \int_0^\infty e_{b\lambda}\, d\lambda = \sigma T^4 \qquad (4.5.1)$$

where σ is the Stefan-Boltzmann constant and is equal to 5.6697 $\times 10^{-8}$ W/m^2 °K^4.

4.6 RADIATION TABLES

Starting with Planck's law [Eq. (4.4.1)] of the spectral distribution of blackbody radiation, Dunkle (1954) has presented a method for simplifying blackbody calculations. Planck's law can be written as

$$e_{b\lambda} = \frac{C_1}{\lambda^5(e^{C_2/\lambda T} - 1)} \qquad (4.6.1)$$

Equation (4.6.1) can be integrated to give the radiation between any wavelength limits. The total emitted from zero to any wavelength λ is given by

$$e_{b,0-\lambda} = \int_0^\lambda e_{b\lambda}\, d\lambda \qquad (4.6.2)$$

Substituting Eq. (4.6.1) into (4.6.2) and noting that by dividing by σT^4, the integral can be made to be only a function of λT, we have

$$\frac{e_{b,0-\lambda T}}{\sigma T^4} = \int_0^{\lambda T} \frac{C_1 d(\lambda T)}{\sigma(\lambda T)^5(e^{C_2/\lambda T} - 1)} \qquad (4.6.3)$$

The values of these integrals have been calculated in convenient wavelength intervals by Sargent (1972) and the results are given in Table 4.6.1. (Note that when the upper limit of integration of Eq. (4.6.3) is ∞, the value of the integral is unity.)

For use on a digital computer, the following polynomial approximations to Eq. (4.6.3) have been given by Pivovonsky and Nagel (1961). For ν greater than or equal to 2:

$$\frac{e_{b,0-\lambda T}}{\sigma T^4} = \frac{15}{\pi^4} \sum_{m=1,2,\ldots} \frac{e^{-m\nu}}{m^4} \left\{ [(m\nu + 3)\ m\nu + 6]\ m\nu + 6 \right\}$$

(4.6.4)

For ν less than 2:

$$\frac{e_{b,0-\lambda T}}{\sigma T^4} = 1 - \frac{15}{\pi^4}\ \nu^3 \left(\frac{1}{3} - \frac{\nu}{8} + \frac{\nu^2}{60} - \frac{\nu^4}{5040} + \frac{\nu^6}{272160} - \frac{\nu^8}{13305600} \right)$$

(4.6.5)

where $\nu = C_2/\lambda T$.

Example 4.6.1 Let us assume that the sun is a blackbody at 5762 °K. (A) What is the wavelength at which the maximum monochromatic emissive power occurs? (B) What is the energy from this 5762 °K source that is in the visible (0.38 to 0.78 μm)?

The value of λT at which the maximum monochromatic emissive power occurs is 2897.8 μm °K. Therefore, the desired wavelength is 2897.8/5762 or 0.503 μm. From Table 4.6.1 the fraction of energy between zero and λT = 0.78 × 5762 = 4494 μm °K is 56.3% and the fraction of the energy between zero and λT = 0.38 × 5762 = 2190 μm °K is 9.9%. Therefore, the fraction of the energy in the visible is 56.3% minus 9.9% or 46.4%.

Note that these numbers compare favorably with the values obtained from the actual distribution of energy from the sun as calculated in Example 1.3.1.

4.7 RADIATION INTENSITY AND FLUX

Thus far, we have considered the radiation leaving a black surface in all directions; however, it is often necessary to describe the directional characteristics of a general radiation field in space. The radiation intensity is used for this purpose and is defined as the energy passing through an imaginary

TABLE 4.6.1 Fraction of Blackbody Radiant Energy Between Zero and λT. From Sargent (1972).

FRACTION OF BLACK-BODY RADIANT ENERGY BETWEEN ZERO AND LAMDA*T

PROPORTIONAL PARTS

LAMDA*TEMP (MICRON*K)	.0	.1	.2	.3	.4
500.	.000000	.000000	.000000	.000000	.000000
600.	.000000	.000000	.000000	.000000	.000000
700.	.000002	.000002	.000003	.000004	.000005
800.	.000016	.000020	.000024	.000028	.000034
900.	.000087	.000101	.000116	.000133	.000152
1000.	.000321	.000360	.000403	.000450	.000501
1100.	.000912	.001001	.001096	.001198	.001308
1200.	.002136	.002305	.002485	.002674	.002875
1300.	.004319	.004603	.004900	.005211	.005536
1400.	.007794	.008225	.008671	.009135	.009614
1500.	.012856	.013459	.014080	.014720	.015378
1600.	.019728	.020520	.021333	.022165	.023017
1700.	.028545	.029537	.030548	.031580	.032631
1800.	.039356	.040546	.041756	.042985	.044234
1900.	.052126	.053507	.054907	.056325	.057761
2000.	.066751	.068310	.069886	.071478	.073087
2100.	.083077	.084797	.086531	.088280	.090043
2200.	.100917	.102776	.104648	.106533	.108430
2300.	.120060	.122038	.124026	.126025	.128034
2400.	.140290	.142365	.144448	.146539	.148638
2500.	.161392	.163542	.165699	.167861	.170030
2600.	.183159	.185364	.187575	.189789	.192008
2700.	.205397	.207640	.209886	.212134	.214385
2800.	.227932	.230195	.232460	.234725	.236992
2900.	.250604	.252873	.255142	.257411	.259679
3000.	.273274	.275535	.277797	.280056	.282314
3100.	.295822	.298066	.300307	.302546	.304783
3200.	.318144	.320360	.322573	.324783	.326989
3300.	.340152	.342333	.344510	.346683	.348852
3400.	.361776	.363915	.366049	.368179	.370304
3500.	.382956	.385048	.387135	.389216	.391293
3600.	.403646	.405686	.407722	.409752	.411776
3700.	.423809	.425795	.427776	.429751	.431721
3800.	.443419	.445349	.447273	.449191	.451104
3900.	.462457	.464329	.466195	.468055	.469909
4000.	.480911	.482724	.484531	.486332	.488127
4100.	.498774	.500527	.502275	.504017	.505753
4200.	.516044	.517739	.519427	.521110	.522787
4300.	.532725	.534361	.535991	.537615	.539234
4400.	.548823	.550401	.551973	.553539	.555100
4500.	.564345	.565866	.567381	.568891	.570395
4600.	.579303	.580758	.582228	.583682	.585131
4700.	.593709	.595120	.596526	.597926	.599321
4800.	.607578	.608936	.610288	.611636	.612978
4900.	.620924	.622230	.623531	.624828	.626119
5000.	.633762	.635018	.636270	.637517	.638759
5100.	.646109	.647317	.648521	.649720	.650914
5200.	.657981	.659143	.660300	.661452	.662600
5300.	.669395	.670511	.671624	.672732	.673835
5400.	.680366	.681440	.682509	.683574	.684635
5500.	.690913	.691944	.692972	.693996	.695016
5600.	.701049	.702041	.703029	.704013	.704993
5700.	.710793	.711746	.712695	.713641	.714583
5800.	.720158	.721074	.721987	.722896	.723801
5900.	.729160	.730041	.730918	.731792	.732662
6000.	.737813	.738660	.739504	.740344	.741181
6100.	.746133	.746947	.747758	.748566	.749370

TABLE 4.6.1 (Continued)

FRACTION OF BLACK-BODY RADIANT ENERGY BETWEEN ZERO AND LAMDA*T

PROPORTIONAL PARTS

LAMDA*TEMP (MICRON*K)	.5	.6	.7	.8	.9
500.	.000000	.000000	.000000	.000000	.000000
600.	.000000	.000001	.000001	.000001	.000001
700.	.000006	.000007	.000009	.000011	.000014
800.	.000040	.000047	.000055	.000065	.000075
900.	.000174	.000197	.000224	.000253	.000285
1000.	.000556	.000617	.000682	.000753	.000829
1100.	.001425	.001550	.001683	.001825	.001976
1200.	.003086	.003309	.003543	.003789	.004048
1300.	.005875	.006229	.006597	.006981	.007380
1400.	.010111	.010625	.011156	.011705	.012271
1500.	.016055	.016751	.017466	.018201	.018954
1600.	.023888	.024780	.025691	.026623	.027574
1700.	.033702	.034793	.035904	.037035	.038186
1800.	.045501	.046788	.048094	.049419	.050763
1900.	.059215	.060687	.062177	.063684	.065209
2000.	.074713	.076354	.078011	.079685	.081373
2100.	.091821	.093613	.095418	.097238	.099071
2200.	.110339	.112260	.114193	.116138	.118094
2300.	.130053	.132082	.134120	.136168	.138225
2400.	.150746	.152860	.154983	.157112	.159249
2500.	.172205	.174385	.176571	.178762	.180958
2600.	.194230	.196457	.198687	.200920	.203157
2700.	.216638	.218893	.221150	.223409	.225669
2800.	.239260	.241528	.243795	.246065	.248334
2900.	.261947	.264214	.266481	.268746	.271011
3000.	.284570	.286824	.289077	.291327	.293576
3100.	.307017	.309248	.311476	.313702	.315924
3200.	.329192	.331391	.333587	.335779	.337968
3300.	.351016	.353177	.355333	.357485	.359633
3400.	.372425	.374541	.376652	.378758	.380860
3500.	.393365	.395431	.397493	.399549	.401600
3600.	.413795	.415809	.417817	.419820	.421817
3700.	.433685	.435643	.437596	.439543	.441484
3800.	.453011	.454912	.456807	.458696	.460579
3900.	.471757	.473600	.475436	.477267	.479092
4000.	.489916	.491700	.493477	.495248	.497014
4100.	.507483	.509207	.510925	.512637	.514344
4200.	.524458	.526123	.527783	.529436	.531084
4300.	.540846	.542453	.544054	.545650	.547239
4400.	.556655	.558204	.559748	.561286	.562818
4500.	.571894	.573387	.574874	.576356	.577832
4600.	.586574	.588012	.599445	.590872	.592293
4700.	.600710	.602094	.603473	.604847	.606215
4800.	.614315	.615647	.616974	.618296	.619612
4900.	.627405	.628686	.629963	.631234	.632500
5000.	.639996	.641228	.642455	.643678	.644896
5100.	.652103	.653288	.654468	.655644	.656815
5200.	.663744	.664883	.666018	.667148	.668273
5300.	.674935	.676030	.677120	.678207	.679289
5400.	.685692	.686744	.687792	.688837	.689877
5500.	.696031	.697043	.698050	.699054	.700054
5600.	.705969	.706942	.707910	.708875	.709836
5700.	.715521	.716456	.717387	.718314	.719238
5800.	.724703	.725601	.726496	.727388	.728275
5900.	.733529	.734393	.735253	.736110	.736963
6000.	.742014	.742844	.743671	.744495	.745316
6100.	.750172	.750970	.751765	.752557	.753346

69

TABLE 4.6.1 (Continued)

FRACTION OF BLACK-BODY RADIANT ENERGY BETWEEN ZERO AND LAMDA*T

PROPORTIONAL PARTS

LAMDA*TEMP (MICRON*K)	.0	.1	.2	.3	.4
5200.	.754132	.754915	.755695	.756472	.757245
6300.	.761825	.762577	.763327	.764074	.764818
6400.	.769222	.769947	.770668	.771386	.772102
6500.	.776338	.777035	.777729	.778420	.779109
6600.	.783184	.783855	.784522	.785197	.785850
6700.	.789772	.790417	.791059	.791699	.792337
6800.	.796111	.796732	.797350	.797966	.798580
6900.	.802213	.802811	.803406	.803999	.804590
7000.	.808098	.808663	.809237	.809808	.810377
7100.	.813745	.814299	.814851	.815401	.815949
7200.	.819194	.819728	.820259	.820789	.821317
7300.	.824443	.824957	.825469	.825980	.826488
7400.	.829500	.829996	.830490	.830982	.831472
7500.	.834375	.834852	.835328	.835803	.836275
7600.	.839074	.839534	.839993	.840450	.840906
7700.	.843604	.844048	.844491	.844931	.845371
7800.	.847973	.848402	.848828	.849254	.849677
7900.	.852188	.852601	.853013	.853423	.853832
8000.	.856254	.856653	.857050	.857446	.857841
8100.	.860179	.860564	.860947	.861329	.861710
8200.	.863967	.864338	.864709	.865078	.865445
8300.	.867624	.867983	.868341	.868697	.869052
8400.	.871156	.871503	.871848	.872192	.872535
8500.	.874568	.874903	.875237	.875569	.875900
8600.	.877865	.878188	.878510	.878832	.879152
8700.	.881050	.881363	.881674	.881985	.882294
8800.	.884129	.884431	.884733	.885033	.885332
8900.	.887106	.887398	.887690	.887980	.888269
9000.	.889985	.890268	.890549	.890830	.891110
9100.	.892770	.893043	.893315	.893587	.893858
9200.	.895463	.895728	.895992	.896255	.896516
9300.	.898070	.898326	.898582	.898836	.899089
9400.	.900594	.900841	.901088	.901335	.901580
9500.	.903036	.903276	.903515	.903754	.903991
9600.	.905402	.905634	.905866	.906097	.906327
9700.	.907692	.907918	.908142	.908366	.908589
9800.	.909912	.910130	.910347	.910564	.910780
9900.	.912062	.912273	.912484	.912694	.912903
10000.	.914146	.916156	.918125	.920024	.921867
11000.	.931832	.933328	.934781	.936194	.937566
12000.	.945037	.946166	.947264	.948334	.949375
13000.	.955075	.955942	.956786	.957610	.958412
14000.	.962831	.963507	.964166	.964809	.965438
15000.	.968913	.969446	.969968	.970478	.970976
16000.	.973745	.974172	.974589	.974999	.975398
17000.	.977630	.977975	.978314	.978645	.978970
18000.	.980799	.981070	.981347	.981619	.981885
19000.	.983380	.983613	.983842	.984067	.984287
20000.	.985528	.987322	.988831	.990111	.991202
30000.	.995263	.995678	.996045	.996372	.996664
40000.	.997839	.998031	.998160	.998278	.998386
50000.	.998874	.998935	.998991	.999044	.999093
60000.	.999324	.999354	.999382	.999409	.999434
70000.	.999557	.999574	.999590	.999604	.999619
80000.	.999690	.999700	.999709	.999718	.999727
90000.	.999771	.999777	.999783	.999789	.999795
100000.	.999823	.999827	.999831	.999835	.999839

TABLE 4.6.1 (Continued)

FRACTION OF BLACK-BODY RADIANT ENERGY BETWEEN ZERO AND LAMDA*T

PROPORTIONAL PARTS

LAMDA*TEMP (MICRON*K)	.5	.6	.7	.8	.9
5200.	.758016	.758784	.759548	.760310	.761069
6300.	.765560	.766298	.767033	.767766	.768496
6400.	.772815	.773525	.774233	.774937	.775639
6500.	.779794	.780478	.781158	.781836	.782512
6600.	.786510	.787167	.787822	.788474	.789124
6700.	.792972	.793604	.794235	.794863	.795488
6800.	.799191	.799800	.800407	.801011	.801613
6900.	.805178	.805765	.806349	.806931	.807511
7000.	.810943	.811508	.812070	.812631	.813189
7100.	.816495	.817039	.817581	.818120	.818658
7200.	.821843	.822367	.822889	.823409	.823927
7300.	.826995	.827500	.828003	.828504	.829003
7400.	.831960	.832447	.832931	.833414	.833895
7500.	.836746	.837215	.837682	.838147	.838611
7600.	.841359	.841812	.842262	.842711	.843158
7700.	.845808	.846244	.846679	.847112	.847543
7800.	.850100	.850520	.850939	.851357	.851773
7900.	.854239	.854645	.855050	.855453	.855854
8000.	.858234	.858626	.859016	.859405	.859793
8100.	.862090	.862468	.862844	.863220	.863594
8200.	.865812	.866177	.866541	.866903	.867264
8300.	.869406	.869758	.870110	.870460	.870809
8400.	.872877	.873218	.873557	.873895	.874232
8500.	.876231	.876560	.876888	.877214	.877540
8600.	.879471	.879789	.880106	.880422	.880736
8700.	.882603	.882910	.883216	.883522	.883826
8800.	.885630	.885927	.886224	.886519	.886813
8900.	.888558	.888845	.889131	.889417	.889701
9000.	.891389	.891667	.891944	.892220	.892495
9100.	.894128	.894397	.894665	.894932	.895198
9200.	.896778	.897038	.897297	.897556	.897813
9300.	.899342	.899594	.899845	.900095	.900345
9400.	.901825	.902069	.902312	.902554	.902795
9500.	.904228	.904464	.904700	.904935	.905168
9600.	.906555	.906785	.907013	.907240	.907467
9700.	.908811	.909032	.909253	.909473	.909693
9800.	.910995	.911210	.911424	.911637	.911850
9900.	.913112	.913320	.913527	.913734	.913940
10000.	.923654	.925388	.927071	.928705	.930292
11000.	.938900	.940197	.941458	.942684	.943876
12000.	.950388	.951376	.952337	.953273	.954186
13000.	.959195	.959959	.960704	.961430	.962139
14000.	.966051	.966651	.967236	.967808	.968367
15000.	.971464	.971940	.972406	.972862	.973308
16000.	.975790	.976174	.976549	.976917	.977277
17000.	.979288	.979600	.979906	.980206	.980500
18000.	.982146	.982403	.982654	.982901	.983143
19000.	.984503	.984716	.984924	.985129	.985330
20000.	.992138	.992946	.993647	.994258	.994793
30000.	.996925	.997159	.997370	.997560	.997733
40000.	.998485	.998576	.998660	.998737	.998808
50000.	.999138	.999180	.999220	.999257	.999291
60000.	.999458	.999480	.999501	.999521	.999539
70000.	.999632	.999645	.999657	.999669	.999680
80000.	.999735	.999743	.999751	.999758	.999765
90000.	.999800	.999805	.999810	.999815	.999819
100000.	.999843	.999846	.999849	.999852	.999855

plane per unit area per unit time and per unit solid angle whose central direction is perpendicular to the imaginary plane. Thus, in Figure 4.7.1, if ΔE represents the energy per unit time passing through ΔA and remaining within $\Delta \omega$, then the intensity is

$$I = \lim_{\substack{\Delta A \\ \Delta \omega}\to 0} \frac{\Delta E}{\Delta A \Delta \omega} \qquad (4.7.1)$$

The intensity I has both a magnitude and a direction and can be considered as a vector quantity. It should be pointed out that for a given imaginary plane in space, we can consider two intensity vectors which are in opposite directions. These two vectors are often distinguished by the symbol I^+ and I^-.

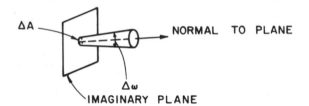

Figure 4.7.1 Schematic of radiation intensity.

The radiation flux is closely related to the intensity and is defined as the energy passing through an imaginary plane per unit area, per unit time and in all directions on one side of the imaginary plane. Note that one difference between intensity and flux is that the differential area for intensity is perpendicular to the direction of propagation, whereas the differential area for flux lies in a plane which forms the base of a hemisphere through which the radiation is passing.

The intensity can be used to determine the flux through any plane. Consider an elemental area ΔA on an imaginary plane covered by a hemisphere of radius r as shown in Figure 4.7.2. The energy per unit time passing through an area $\Delta A'$ on the surface of the hemisphere from the area ΔA is approximately equal to

$$\Delta Q = I \Delta A \cos \theta \frac{\Delta A'}{r^2} \qquad (4.7.2)$$

where $\Delta A'/r^2$ is the solid angle between ΔA and $\Delta A'$ and $\Delta A \cos \theta$ is the area in the direction of the intensity vector. The energy flux per unit solid angle in the θ, ϕ direction can then be defined as

$$\Delta q = \lim_{\Delta A \to 0} \frac{\Delta Q}{\Delta A} = I \cos \theta \frac{\Delta A'}{r^2} \qquad (4.7.3)$$

The radiation flux is then found by summing all Δq over the hemisphere. The sphere incremental area can be expressed in terms of the angles θ and ϕ so that

$$q = \int_0^{2\pi} \int_0^{\pi/2} I \cos \theta \sin \theta \, d\theta \, d\phi \qquad (4.7.4)$$

It is convenient to define $\mu = \cos \theta$. Then we can write

$$q = \int_0^{2\pi} \int_0^1 I \mu \, d\mu \, d\phi \qquad (4.7.5)$$

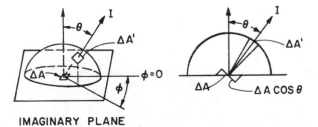

IMAGINARY PLANE

Figure 4.7.2 Schematic of radiation flux.

Two important points concerning the radiation flux must be remembered. First, the radiation flux is, in general, a function of the orientation of the chosen imaginary plane. Second, the radiation flux will have two values corresponding to each of the two possible directions of the normal to the imaginary plane. When it is necessary to emphasize which of the two possible values of the radiation flux is being considered, the

superscript + or - can be used along with a definition of the
positive and negative direction.

Thus far, we have defined radiation flux and intensity at
a general location in space. When it is desired to find the
heat transfer between surfaces in a vacuum, or at least in a
radiative nonparticipating media such as a thin air space, the
most useful values of radiative flux and intensity occur at the
surfaces. For the special case of a surface that has I independent of direction, the integration of Eq. (4.7.5) yields

$$q = \pi I \qquad\qquad (4.7.6)$$

Surfaces that have the intensity equal to a constant are called
either *Lambertonian* or diffuse surfaces. A blackbody emits in
a diffuse manner and therefore the blackbody emissive power is
related to the blackbody intensity by

$$e_b = \pi I_b \qquad\qquad (4.7.7)$$

The foregoing equations were written for total radiation,
but apply equally well to monochromatic radiation. For example,
Eq. (4.7.7) could be written in terms of a particular wavelength

$$e_{b\lambda} = \pi I_{b\lambda} \qquad\qquad (4.7.8)$$

4.8 INFRARED RADIATION HEAT TRANSFER BETWEEN GRAY SURFACES

The general case of infrared radiation heat transfer between
many gray surfaces having different temperatures is treated in
a number of textbooks [e.g., Hottel and Sarofim (1967), Siegel
and Howell (1972)]. The various methods all make the same basic
assumptions which, for each surface, can be summarized as

1. The surface is gray. (All radiation properties are
   independent of wavelength.)
2. The surface is diffuse or specular-diffuse. (See section 5.3.)
3. The surface temperature is uniform.
4. The incident energy over the surface is uniform.

Beckman (1971) also utilized these basic assumptions and
defined a total exchange factor between pairs of surfaces of an
N surface enclosure such that the net heat flux to a typical

surface, i, is*

$$Q_i = \sum_{j=1}^{N} \epsilon_i \epsilon_j A_i \hat{F}_{ij} \sigma(T_j^4 - T_i^4) \qquad (4.8.1)$$

The total exchange factors (i.e., the $\hat{F}_{ij}$'s) between surface i and j are found from the matrix equation

$$[\hat{F}_{ij}] = [\delta_{ij} - \rho_j E_{ij}]^{-1}[E_{ij}] \qquad (4.8.2)$$

The specular exchange factors (i.e., the E_{ij}'s) of this equation account for radiation going from surface i to surface j directly and by all possible specular (mirror-like) reflections. Methods for calculating E_{ij} are given in most advanced radiation heat transfer textbooks. When the surfaces of the enclosure do not specularly reflect radiation, the specular exchange factors of Eq. (4.8.2) reduce to the usual view factor (configuration factor) defined in all elementary heat transfer textbooks.

The majority of heat transfer problems in solar energy applications involve radiation between two surfaces. The solution of Eqs. (4.8.1) and (4.8.2) for diffuse surface with $N = 2$ is

$$Q_1 = -Q_2 = \frac{\sigma(T_2^4 - T_1^4)}{(1 - \epsilon_1)/\epsilon_1 A_1 + 1/A_1 F_{12} + (1 - \epsilon_2)/\epsilon_2 A_2} \qquad (4.8.3)$$

Two special cases of Eq. (4.8.3) are of particular interest. For radiation between two infinite parallel plates the area A_1 and A_2 are equal and the view factor F_{12} is unity. Under these conditions, Eq. (4.8.3) becomes

$$\frac{Q}{A} = \frac{\sigma(T_2^4 - T_1^4)}{1/\epsilon_1 + 1/\epsilon_2 - 1} \qquad (4.8.4)$$

The second special case is for a small convex object (surface 1) surrounded by a large enclosure (surface 2). Under these conditions, the area ratio A_1/A_2 approaches zero, the view factor F_{12} is unity, and Eq. (4.8.3) becomes

*The emittance, ϵ, is defined by Eq. (5.1.8).

$$Q_1 = \varepsilon_1 A_1 \sigma (T_2^4 - T_1^4) \qquad (4.8.5)$$

This result is independent of the surface properties of the large enclosure since virtually none of the radiation leaving the small object is reflected back from the large enclosure. In other words, the large enclosure absorbs all radiation from the small object and acts like a blackbody. Equation (4.8.5) also applies in the case of a flat plate radiating to the sky, as is found with a collector cover radiating to the surroundings.

4.9 SKY RADIATION

In order to predict the performance of solar collectors it will be necessary to evaluate the radiation exchange between a surface and the sky. The sky can be considered as a blackbody at some equivalent sky temperature so that the actual net radiation between a flat plate facing the sky and the sky is given by Eq. (4.8.5). The net radiation to a surface with emittance ε and temperature T is thus found from

$$Q = \varepsilon A \sigma (T_{\text{sky}}^4 - T^4) \qquad (4.9.1)$$

The equivalent blackbody sky temperature of Eq. (4.9.1) accounts for the facts that the atmosphere is not at a uniform temperature and that the atmosphere radiates only in certain wavelength bands. The atmosphere is essentially transparent in the wavelength region from 8 to 14 μm while outside the "window" the atmosphere has radiating bands covering much of the infrared spectrum. Several relations for clear skies have been proposed to relate T_{sky} to other measured meteorological variables. Brunt (1932) and Bliss (1961) relate the effective sky temperature to water vapor content of the air and/or air temperature. Swinbank (1963) relates sky temperature to the local air temperature in the simple relationship

$$T_{\text{sky}} = 0.0552 \, T_{\text{air}}^{1.5} \qquad (4.9.2)$$

where T_{sky} and T_{air} are both in degrees Kelvin. Whillier (1967) uses an even simpler expression and substracts 6 °C from the air temperature.

$$T_{\text{sky}} = T_{\text{air}} - 6 \qquad (4.9.3)$$

Equations (4.9.2) and (4.9.3) give identical results when the air temperature is 308 °K. This seems to be a reasonable summer condition but Whillier's 6 °C should probably be raised to about 20 °C for winter operation.
It is not at all clear which relationship should be used. The influences of clouds and ground, if the collector is tilted and can see the ground, have not been included in either expression. It is likely that both the clouds and the ground will tend to increase the effective sky temperature over that for a clear sky. In any event, it is fortunate that it does not make much difference which expression is used in evaluating collector performance (see section 7.4).

4.10 RADIATION HEAT TRANSFER COEFFICIENT

In order to retain the simplicity of linear equations it is convenient to define a radiation heat transfer coefficient. The heat transfer by radiation between two arbitrary surfaces is found from Eq. (4.8.3). If we define a heat transfer coefficient so that the radiation between the two surfaces is given by

$$Q = A_1 h_r (T_2 - T_1) \qquad (4.10.1)$$

then it is clear that

$F_{12} \rightarrow$ page 75

$$h_r = \frac{\sigma(T_2^2 + T_1^2)(T_2 + T_1)}{(1 - \varepsilon_1)/\varepsilon_1 + 1/F_{12} + (1 - \varepsilon_2)A_1/\varepsilon_2 A_2}$$

$$(4.10.2)$$

It is important to remember that if the areas A_1 and A_2 are not equal, then the numerical value of h_r depends upon whether it is to be used with A_1 or with A_2.
The numerator of Eq. (4.10.2) can be expressed as $4\sigma \bar{T}^3$. Since T_1 and T_2 are often close together, it is not difficult to estimate $\bar{T}$ without actually knowing T_1 and T_2. If $\bar{T}$ is known, the equations of radiation heat transfer are reduced to linear equations that can be easily solved along with the linear equations of conduction and convection. If more accuracy is needed, a second or third iteration may be required.

4.11 NATURAL CONVECTION BETWEEN PARALLEL FLAT PLATES

The rate of heat transfer between two plates inclined at some

angle to the horizon is of obvious importance in the performance
of flat-plate collectors. Tabor (1958) has examined the published results of a number of investigations and concluded that
the most reliable data for use in solar collector calculations
is contained in a report which was published by the Home Finance
Agency in Washington, D.C. (1954). Tabor replotted the data of
this report in the form of Nusselt number (Nu) as a function of
Grashof (Gr) number for five different inclines and/or directions
of heat flow. An adaptation of Tabor's graph is shown in Figure
4.11.1. The Nusselt number and Grashof numbers of Figure 4.11.1
are given by

$$Nu = \frac{hL}{k} \qquad\qquad (4.11.1)$$

$$Gr = \frac{g\beta\Delta T L^3}{\nu^2} \qquad\qquad (4.11.2)$$

where h = heat transfer coefficient;
        L = plate spacing;
        k = thermal conductivity;
        g = gravitational constant;
        β = volumetric coefficient of expansion of air
            (1/T for an ideal gas);
        ΔT = temperature difference between plates;
        ν = kinematic viscosity.

and all fluid properties are evaluated at the mean air temperature. Note that for parallel plates the Nusselt number is the
ratio of a pure conduction resistance to a convection resistance
[i.e., Nu = $(L/K)/(1/h)$] so that a Nusselt number of unity represents pure conduction.
    In addition to the Nusselt number, Figure 4.11.1 has another
scale on the ordinate giving the value of the heat transfer coefficient times the plate spacing for a mean temperature of 10 °C.
The scale of this ordinate is not dimensionless but is W cm/m² °C.
For temperatures other than 10 °C, a factor F_2 has been plotted
as a function of temperature in Figure 4.11.2. This factor is
the ratio of the thermal conductivity of air at 10 °C to that
at any other temperature. Thus, to find hl^* at any temperature
other than 10 °C, it is only necessary to divide $F_2 h l$ as read
from the chart by F_2 at the appropriate temperature.
    The abscissa also has an extra scale for 10 °C; F_1 $\Delta T l^3$ is in
cm³ °C. To find $\Delta T l^3$ at temperatures other than 10 °C, it is
only necessary to divide $F_1 \Delta T l^3$ by F_1, where F_1 is the ratio of
β/ν^2 at the desired temperature to β/ν^2 at 10 °C; F_1 is also plotted on
Figure 4.11.2.

*The lower case letter l is used to indicate that the units are
centimeters instead of meters as indicated by the capital L.

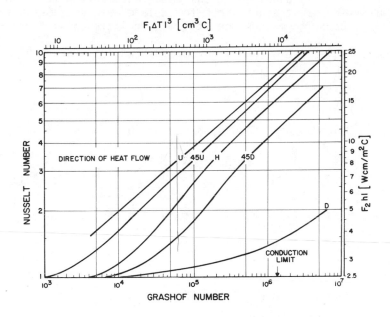

Figure 4.11.1 Nusselt number as a function of Grashof number for free convection heat transfer between parallel planes. [Adapted from Tabor, (1958).]

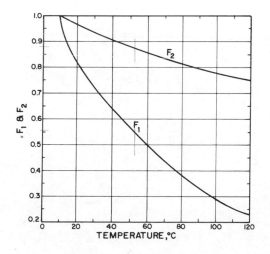

Figure 4.11.2 Air property corrections F_1 and F_2 for Figure 4.11.1.

The Nusselt-Grashof relations shown on Figure 4.11.1 can be represented as straight lines for a large range of Grashof numbers. Tabor recommends the following relationships for air:

horizontal planes, heat flow upwards and $10^4 < Gr < 10^7$

$$Nu = 0.152 \, (Gr)^{0.281}$$

(4.11.3)

45° planes, heat flow upwards and $10^4 < Gr < 10^7$

$$Nu = 0.093 \, (Gr)^{0.310}$$

(4.11.4)

vertical planes, $1.5 \times 10^5 < Gr < 10^7$

$$Nu = 0.062 \, (Gr)^{0.327}$$

(4.11.5)

vertical planes, $1.5 \times 10^4 < Gr < 1.5 \times 10^5$

$$Nu = 0.033 \, (Gr)^{0.381}$$

(4.11.6)

The above equations can be conveniently expressed in dimensional form for temperatures of 10 °C as

$$h_{10} = 1.613 \, \frac{\Delta T^{0.281}}{l^{0.157}}$$

(4.11.3a)

$$h_{10} = 1.14 \, \frac{\Delta T^{0.310}}{l^{0.070}}$$

(4.11.4a)

$$h_{10} = 0.82 \, \frac{\Delta T^{0.327}}{l^{0.019}}$$

(4.11.5a)

$$h_{10} = 0.57 \, \frac{\Delta T^{0.381}}{l^{-0.143}}$$

(4.11.6a)

where h is in W/m^2 °C, ΔT is in degrees Celsius, and l is the plate spacing in centimeters. For temperatures other than 10 °C, the following relation can be used to correct for property variations:

$$\frac{h_T}{h_{10}} = 1 - 0.0018(\bar{T} - 10) \qquad (4.11.7)$$

where $\bar{T}$ is the average temperature between the two plates in degrees Celsius

In a more recent experimental investigation using liquids between parallel planes, Dropkin and Somerscales (1965) presented correlations which give somewhat lower heat transfer coefficients. Their results can be conveniently represented for angles of tilt from 0° to 90° and Gr greater than 2×10^5 as

$$\text{Nu} = \left[0.069 - 0.020 \left(\tfrac{s}{90}\right)\right](\text{GrPr})^{\frac{1}{3}} (\text{Pr})^{0.074} \qquad (4.11.8)$$

where s is the tilt angle in degrees*.

Since most solar collectors are concerned with air, and the Prandtl number of air is 0.7 and nearly independent of temperature, Eq. (4.11.8) becomes

$$\text{Nu} = \left[0.060 - 0.017 \left(\tfrac{s}{90}\right)\right]\text{Gr}^{\frac{1}{3}} \qquad (4.11.9)$$

Example 4.11.1 Find the heat transfer coefficient between two parallel plates separated by 2.5 cm with a 45° tilt. The lower plate is maintained at 70 °C and the upper plate at 50 °C.

From Figure 4.11.2, at a mean temperature of 60 °C, F_1 = 0.49 and F_2 = 0.86. Therefore, $F_1\Delta T l^3$ = 0.49 × 20 × 2.5³ = 153 °C cm³. From the curve marked 45U on Figure 4.11.1, $F_2 h_c l$ = 5.3. Finally, h_c = 5.3/(0.86 × 2.5) = 2.4 W/m² °C.

4.12 CONVECTION SUPPRESSION USING HONEYCOMBS

One of the objectives in designing solar collectors is to reduce the heat loss through the covers. This desire has led to convection suppression studies by Hollands (1965), Edwards (1969), and Charters and Peterson (1972). In these studies the space

*Grashof numbers less than 2×10^5 are often encountered in solar collectors.

between two plates, with one plate heated, is filled with a
honeycomb material to suppress the onset of fluid motion. Without fluid motion the heat transfer between the plates is by conduction and radiation.

Hollands and Edwards both studied horizontal systems with
heat flow upward and concluded that the onset of convection,
with its large increase in heat transfer rates, could be sup-
pressed by using honeycomb materials. However, Charters and
Peterson experimentally investigated inclined systems and observed little or no convection suppression when the angle of
inclination was only a few degrees from horizontal. Since most
flat-plate solar collectors do not operate in a horizontal position, the use of honeycombs may actually prove to be a disadvantage since the honeycombs may significantly reduce the transmission of solar radiation. Further experimental studies are
needed to clarify the situation.

4.13 *FORCED CONVECTION BETWEEN PARALLEL FLAT PLATES*

In the study of solar air heaters, it is necessary to know the
convection heat transfer coefficient between two flat plates
for laminar and turbulent flow. Kays (1966) gives a complete
description of both constant heat flux and constant wall temperature conditions. For the laminar fully developed hydro-
dynamic and thermal problem, with one surface insulated, Kays
gives Nusselt numbers of 5.4 and 4.9 for constant heat flux and
constant temperature conditions, respectively.

For air, the following correlation for fully developed turbulent flow between flat plates with one side heated can be derived from Kays' data:

$$Nu = 0.0158 \ Re^{0.8} \qquad (4.13.1)$$

Whenever the ratio of flow length to hydraulic diameter is on
the order of ten or less, consideration should be given to the
development of the boundary layers which results in an increase
in the Nusselt number.

Tan and Charters (1970) have experimentally studied flow
of air between parallel plates with small aspect ratios for use
in solar air heaters. Their results give higher heat transfer
coefficients by about 10% than those given by Kays with an infinite aspect ratio.

4.14 *MISCELLANEOUS HEAT TRANSFER RELATIONS*

Heat transfer coefficients for common geometries are given in many heat transfer textbooks [e.g., Kreith (1973), Kays (1966), and McAdams (1954)]. For fully developed turbulent flow of gases inside tubes, Kays recommends

$$Nu = 0.022 \ Re^{0.8} \ Pr^{0.6} \tag{4.14.1}$$

and for liquids with the Prandtl number between 1.0 and 20.

$$Nu = 0.0155 \ Re^{0.83} \ Pr^{0.5} \tag{4.14.2}$$

For laminar flow inside short tubes, Kreith gives the following equation which is useful since the pipes in solar systems are often short, and fully developed conditions may not exist:

$$Nu = \frac{Re \ Pr \ D}{4L} \ \ln \left[1 - \frac{2.654}{Pr^{0.167} \ (Re \ Pr \ D/L)^{0.5}} \right]^{-1} \tag{4.14.3}$$

This result is valid for flow in tubes when L/D is less than 0.0048 Re, and in flat ducts when L/D is less than 0.0021 Re. The heat loss from flat plates exposed to outside winds are found from a dimensional expression given by McAdams which relates the heat transfer coefficient in $W/m^2 \ °C$ to the wind speed in m/s.

$$h_{wind} = 5.7 + 3.8 \ V \tag{4.14.4}$$

REFERENCES

Beckman, W. A., Solar Energy, *13*, 3 (1971). "The Solution of Heat Transfer Problems on a Digital Computer."

Bliss, R. W., Solar Energy, *5*, 103 (1961). "Atmospheric Radiation Near the Surface of the Ground."

Brunt, D., Quart. J. Roy. Meteorol. Soc., *58*, 389-420,(1932). "Notes on Radiation in the Atmosphere."

Charters, W. W. S. and Peterson, L. F., Solar Energy, *13*, 4 (1972). "Free Convection Suppression Using Honeycomb Celluar Materials."

Dropkin, D. and Somerscales, E., ASME, J. Heat Transfer, *87*, 77 (1965). "Heat Transfer by Natural Convection in Liquids Confined by Two Parallel Plates Which are Inclined at Various Angles with Respect to the Horizontal."

Dunkle, R. V., Trans. ASME, *76* (1954). "Thermal Radiation Tables and Applications."

Edwards, D. K., Trans. ASME, J. Heat Transfer, *91*, 145 (1969). "Suppression of Cellular Convection by Lateral Walls."

Hollands, K. G. T., Solar Energy, *9*, 159 (1965). "Honeycomb Devices in Flat-Plate Solar Collectors."

Home Finance Agency Report #32, U.S. Government Printing Office (1954). "The Thermal Insulating Value of Airspaces."

Hottel, H. C. and Sarofim, A. F., <u>Radiative Transfer</u>, New York, McGraw-Hill, 1967.

Kays, W. M., <u>Convective Heat and Mass Transfer</u>, New York, McGraw-Hill, 1966.

Kreith, F., <u>Principles of Heat Transfer</u>, 3rd ed., Scranton, Pa., International Textbook Co., 1973.

McAdams, W. C., <u>Heat Transmission</u>, 3rd ed., New York, McGraw-Hill, 1954.

Pivovonsky, M. and Nagel, M. R., <u>Tables of Blackbody Radiation Properties</u>, New York, Macmillan, 1961.

Richtmyer, F. K. and Kennard, E. H., <u>Introduction to Modern Physics</u>, 4th ed., New York, McGraw-Hill, 1947.

Sargent, S. L., Bull. Am. Meteorol. Soc., *53*, 360 (April 1972). "A Compact Table of Blackbody Radiation Functions."

Siegel, R. and Howell, J. R., <u>Thermal Radiation Heat Transfer</u>, New York, McGraw-Hill, 1972.

Swinbank, W. C., Quart. J. Roy. Meteorol. Soc., *89* (1963). "Long-Wave Radiation From Clear Skies."

Tan, H. M. and Charters, W. W. S., Solar Energy, *13*, 121 (1970). "Experimental Investigation of Forced-Convection Heat Transfer."

Tabor, H., Bull. Res. Coun. Israel, *6C*, 155 (1958). "Radiation Convection and Conduction Coefficients in Solar Collectors."

Whillier, A., Low Temperature Engineering Applications of Solar Energy, New York, ASHRAE, 1967. "Design Factors Influencing Solar Collectors."

5. RADIATION CHARACTERISTICS OF OPAQUE MATERIALS

In this chapter, a large number of radiation surface characteristics will be defined. The names used for these characteristics were chosen as the most descriptive of the many names
found in the literature. In many cases, the names will seem to
be very cumbersome, but they are necessary to distinguish one
characteristic from another. For example, it will be necessary
to define both a monochromatic angular-hemispherical reflectance
and a monochromatic hemispherical-angular reflectance. Under
certain circumstances, these two quantities are identical, but
in general they are different; consequently, it is necessary to
distinguish them by using separate names.
   Both the names and the symbols should be an aid for understanding the significance of the particular characteristic.
Thus, the monochromatic directional absorptance, $\alpha_\lambda(\mu,\phi)$, is
the fraction of the incident energy from the direction μ,ϕ at
the wavelength λ that is absorbed. The directional absorptance,
$\alpha(\mu,\phi)$, includes all wavelengths and the hemispherical absorptance, α, includes all directions as well as all wavelengths.
We will also have a monochromatic hemispherical absorptance, α_λ,
which is the fraction of the energy incident from all directions,
but at one wavelength, that is absorbed. Thus, by careful study
of the name, the definition should be clear.

5.1 ABSORPTANCE AND EMITTANCE

The monochromatic directional absorptance is a property of a
surface and is defined as the fraction of the incident radiation
of wavelength λ from the direction μ,ϕ (μ is the cosine of the
polar angle and ϕ is the azimuthal angle) that is absorbed by
the surface. In equation form

$$\alpha_\lambda(\mu,\phi) = \frac{I_{\lambda,a}(\mu,\phi)}{I_{\lambda,i}(\mu,\phi)} \qquad (5.1.1)$$

where the subscripts a and i represent absorbed and incident.

The fraction of all the radiation (i.e., over all wavelengths) from the direction μ,ϕ that is absorbed by a surface is called the directional absorptance and is defined by the following equation:

$$\alpha(\mu,\phi) = \frac{\int_0^\infty \alpha_\lambda(\mu,\phi)I_{\lambda,i}(\mu,\phi)\ d\lambda}{\int_0^\infty I_{\lambda,i}(\mu,\phi)\ d\lambda} = \frac{1}{I_i(\mu,\phi)} \int_0^\infty \alpha_\lambda(\mu,\phi)I_{\lambda,i}(\mu,\phi)\ d\lambda$$

$$(5.1.2)$$

Unlike the monochromatic directional absorptance, the directional absorptance is not a property but a function of the wavelength distribution of the incident radiation.

The monochromatic directional emittance of a surface is defined as the ratio of the monochromatic intensity emitted by a surface in a particular direction to the monochromatic intensity that would be emitted by a blackbody at the same temperature.

$$\epsilon_\lambda(\mu,\phi) = \frac{I_\lambda(\mu,\phi)}{I_{b\lambda}} \qquad (5.1.3)$$

The monochromatic directional emittance is a property of a surface, as is the directional emittance, defined by

$$\epsilon(\mu,\phi) = \frac{\int_0^\infty \epsilon_\lambda(\mu,\phi)I_{b\lambda}\ d\lambda}{\int_0^\infty I_{b\lambda}\ d\lambda} = \frac{1}{I_b} \int_0^\infty \epsilon_\lambda(\mu,\phi)I_{b\lambda}\ d\lambda$$

$$(5.1.4)$$

In words, the directional emittance is defined as the ratio of the emitted total intensity in the direction μ,ϕ to the blackbody intensity. The reason that $\alpha(\mu,\phi)$ is not a property whereas $\epsilon(\mu,\phi)$ is a property is that the definition of $\alpha(\mu,\phi)$ contains the unspecified function $I_{\lambda,i}(\mu,\phi)$ but the intensity $I_{b\lambda}$ in the definition of $\epsilon(\mu,\phi)$ is specified when the surface temperature is known. An important point to remember is that these four quantities and the four to follow are all functions of surface conditions such as roughness, temperature, cleanliness, and so on.

From the definitions of the directional absorptance and emittance of a surface, the corresponding hemispherical properties can be defined. The monochromatic hemispherical absorptance and emittance are obtained by integrating over the enclosing hemisphere as was done in section 4.7.

$$\alpha_\lambda = \frac{\int_0^{2\pi} \int_0^1 \alpha_\lambda(\mu,\phi) I_{\lambda i}(\mu,\phi)\mu\ d\mu\ d\phi}{\int_0^{2\pi} \int_0^1 I_{\lambda i}\mu\ d\mu\ d\phi} \qquad (5.1.5)$$

$$\varepsilon_\lambda = \frac{\int_0^{2\pi} \int_0^1 \varepsilon_\lambda(\mu,\phi) I_{b\lambda}\mu\ d\mu\ d\phi}{\int_0^{2\pi} \int_0^1 I_{b\lambda}\mu\ d\mu\ d\phi} = \frac{1}{\pi} \int_0^{2\pi} \int_0^1 \varepsilon_\lambda(\mu,\phi)\mu\ d\mu\ d\phi$$

$$(5.1.6)$$

The monochromatic hemispherical emittance is seen to be a property but the monochromatic hemispherical absorptance is not a property but a function of the incident intensity.

The hemispherical absorptance and emittance are obtained by integrating over all wavelengths and are defined by

$$\alpha = \frac{\int_0^\infty \int_0^{2\pi} \int_0^1 \alpha_\lambda(\mu,\phi) I_{\lambda i}(\mu,\phi)\mu\ d\mu\ d\phi\ d\lambda}{\int_0^\infty \int_0^{2\pi} \int_0^1 I_{\lambda i}(\mu,\phi)\mu\ d\mu\ d\phi\ d\lambda} \qquad (5.1.7)$$

$$\varepsilon = \frac{\int_0^\infty \int_0^{2\pi} \int_0^1 \varepsilon_\lambda(\mu,\phi) I_{b\lambda}\mu\ d\mu\ d\phi\ d\lambda}{\int_0^\infty \int_0^{2\pi} \int_0^1 I_{b\lambda}\mu\ d\mu\ d\phi\ d\lambda} = \frac{1}{e_b} \int_0^\infty \varepsilon_\lambda e_{b\lambda}\ d\lambda$$

$$(5.1.8)$$

Again the absorptance, in this case the hemispherical absorptance, is a function of the incident intensity whereas the hemispherical emittance is a surface property.

5.2 KIRCHOFF'S LAW

A complete proof of Kirchoff's law is beyond the scope of this book [see Siegel and Howell, (1972) for a complete discussion]. However, a satisfactory understanding can be obtained without a proof. Consider an evacuated isothermal enclosure at temperature T. If the enclosure is isolated from the surroundings, then the enclosure and any substance within the enclosure will be in thermodynamic equilibrium. In addition, the radiation field within the enclosure must be homogeneous and isotropic. If this were not so, we could have a directed flow of radiant energy at some location within the enclosure; but this is impossible since we could then extract work from an isolated and isothermal system.

If we now consider an arbitrary body within the enclosure, the body must absorb the same amount of energy as it emits. An energy balance on an element of the surface of the arbitrary body yields

$$\alpha q = \varepsilon e_b \tag{5.2.1}$$

If we place a second body with different surface properties in the enclosure, the same energy balance must apply and the ratio q/e_b must be a constant.

$$\frac{q}{e_b} = \frac{\varepsilon_1}{\alpha_1} = \frac{\varepsilon_2}{\alpha_2} \tag{5.2.2}$$

Since this must also apply to a blackbody in which $\varepsilon = 1$, the ratio of ε to α for any body in thermal equilibrium must be equal to unity. Therefore, for conditions of thermal equilibrium

$$\varepsilon = \alpha \tag{5.2.3}$$

It must be remembered that α is not a property, and since this equation was developed for the condition of thermal equilibrium, it may not be valid if the incident radiation comes from a source at a different temperature (e.g., if the source of radiation is the sun). This distinction is very important in the performance of solar collectors.

Equation (5.2.3) is sometimes referred to as Kirchoff's law, but his law is much more general. Within an enclosure the radiant flux is everywhere uniform and isotropic. The absorptance of a surface within the enclosure is then given by Eq. (5.1.7) with $I_{\lambda,i}(\mu,\phi)$ replaced by $I_{b\lambda}$, and the emittance is Eq. (5.1.8). Since the hemispherical absorptance and emittance are equal under conditions of thermal equilibrium, we can equate Eqs. (5.1.7) and (5.1.8) to obtain

$$\int_0^\infty I_{b\lambda} \int_0^{2\pi} \int_0^1 [\alpha_\lambda(\mu,\phi) - \varepsilon_\lambda(\mu,\phi)]\mu \; d\mu \; d\phi \; d\lambda = 0$$

$$(5.2.4)$$

It is mathematically possible to have this integral equal to zero without $\alpha_\lambda(\mu,\phi)$ being identical to $\varepsilon_\lambda(\mu,\phi)$, but this is a very unlikely situation in view of the very irregular behavior of $\alpha\lambda(\mu,\phi)$ exhibited by some substances. Thus we can say, with a high degree of confidence, that

$$\alpha_\lambda(\mu,\phi) = \varepsilon_\lambda(\mu,\phi) \qquad\qquad (5.2.5)$$

This result is true for all conditions, not just thermal equilibrium, since both $\alpha_\lambda(\mu,\phi)$ and $\varepsilon_\lambda(\mu,\phi)$ are properties, and it is the most general form of Kirchoff's law*.

5.3   REFLECTION FROM SURFACES

Consider the spatial distribution of radiation reflected by a surface. When the incident radiation is in the form of a narrow pencil (a small solid angle), two limiting distributions of the reflected radiation exist. These two cases are called *specular* and *diffuse*. Specular reflection is identical to reflection from a mirror, that is, the incident polar angle is equal to the reflected polar angle and the azimuthal angles differ by 180°. On the other hand, diffuse reflection obliterates all directional characteristics of the incident radiation by distributing the radiation uniformly in all directions. In actual practice, the reflection from a surface is neither specular nor diffuse but is a very complicated phenomenon. The general case along with the two limiting situations is shown in Figure 5.3.1.

In general, the magnitude of the reflected intensity in a particular direction for a given surface is a function of the wavelength and the spatial distribution of the incident radiation. The biangular reflectance or reflection function is used to relate the intensity of reflected radiation in a particular direction by the following equation:

*Kirchoff's law actually applies to each component of polarization and not to the sum of the two components as implied by Eq. (5.2.5). The difference is minor in most applications.

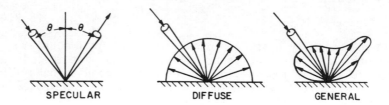

SPECULAR   DIFFUSE   GENERAL

Figure 5.3.1   Reflection from a surface.

$$\rho_\lambda(\mu_r, \phi_r, \mu_i, \phi_i) = \frac{\pi I_{\lambda,r}(\mu_r, \phi_r)}{I_{\lambda,i}\mu_i\Delta\omega_i}$$
(5.3.1)

The numerator is π times the intensity reflected in the direction μ_r, ϕ_r when an energy flux of amount $I_{\lambda,i}\mu_i\Delta\omega_i$ is incident on the surface from the direction μ_i, ϕ_i. The π has been included so that the numerator "looks like" an energy flux. The physical situation is shown schematically in Figure 5.3.2.

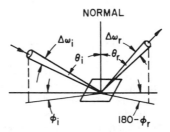

Figure 5.3.2   Coordinate system for the reflection function.

Since the energy incident in the solid angle $\Delta\omega_i$ may be reflected in all directions, the reflected intensity in the direction μ_r, ϕ_r will be of infinitesimal size compared to the incident intensity. By multiplying the incident intensity by its solid angle (which must be finite in any real experiment) and the cosine of the polar angle, we obtain the incident radiant flux which will have values on the same order of magnitude as the reflected intensity. The biangular reflectance can have numerical values between zero and infinity; its values do not lie only between zero and one.

From an experimental point of view, it is not practical to use the scheme depicted in Figure 5.3.2 since all the radiation quantities would be extremely small. An equivalent experiment is to irradiate the surface with a nearly monodirectional flux (i.e., with a small solid angle $\Delta\omega_i$) as shown in Figure 5.3.3.

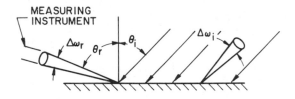

Figure 5.3.3 Schematic representation of an experiment for measuring the reflection function.

The reflected energy in each direction is measured. This measured energy divided by the measurement instruments solid angle ($\Delta\omega_r$) will be approximately equal to the reflected intensity. The incident flux will be on the same order and can be easily measured.

Two types of hemispherical reflectances exist. The angular-hemispherical reflectance is found when a narrow pencil of radiation is incident on a surface and all the reflected radiation is collected. The hemispherical-angular reflectance results from collecting reflected radiation in a particular direction when the surface is irradiated from all directions.

The monochromatic angular-hemispherical reflectance will be designated by $\rho_\lambda(\mu_i,\phi_i)$ where the subscript i indicates that the incident radiation has a specified direction. This reflectance is defined as the ratio of the monochromatic radiant energy reflected in all directions to the incident radiant flux within small solid angle $\Delta\omega_i$. The incident energy ($I_{\lambda,i}\mu_i\Delta\omega_i$) that is reflected in all directions can be found using the reflection function:

$$q_{\lambda,r} = \frac{1}{\pi} \int_0^{2\pi} \int_0^1 \rho_\lambda(\mu_r,\phi_r,\mu_i,\phi_i)I_{\lambda,i}\mu_i\Delta\omega_i\mu_r \, d\mu_r \, d\phi_r$$

$$(5.3.2)$$

The monochromatic angular-hemispherical reflectance can then be expressed as

$$\rho_\lambda(\mu_i,\phi_i) = \frac{q_{\lambda,r}}{I_{\lambda,i}\mu_i\Delta\omega_i} = \frac{1}{\pi} \int_0^{2\pi} \int_0^1 \rho_\lambda(\mu_r,\phi_r,\mu_i\phi_i)\mu_r \, d\mu_r \, d\phi_r$$

$$(5.3.3)$$

Examination of Eq. (5.3.3) shows that $\rho_\lambda(\mu_i, \phi_i)$ is a property of the surface. The angular-hemispherical reflectance $[\rho(\mu_i, \phi_i)]$ can be found by integrating the incident and reflected fluxes over all wavelengths, but it is not a property.

The monochromatic hemispherical-angular reflectance is defined as the ratio of the reflected monochromatic intensity in the direction μ_r, ϕ_r to the monochromatic energy from all directions divided by π (which then "looks like" an intensity).

The incident energy can be written in terms of the incident intensity integrated over the hemisphere:

$$q_{\lambda, i} = \int_0^{2\pi} \int_0^1 I_{\lambda, i} \mu_i \, d\mu_i \, d\phi_i \qquad (5.3.4)$$

and the monochromatic hemispherical-angular reflectance is then

$$\rho_\lambda(\mu_r, \phi_r) = \frac{I_{\lambda, r}(\mu_r, \phi_r)}{q_{\lambda i}/\pi} \qquad (5.3.5)$$

where the subscripts r in $\rho_\lambda(\mu_r, \phi_r)$ are used to specify the reflected radiation as being in a specified direction. In terms of the reflectance function, Eq. (5.3.3) can be written as

$$\rho_\lambda(\mu_r, \phi_r) = \frac{\int_0^{2\pi} \int_0^1 \rho_\lambda(\mu_r, \phi_r, \mu_i, \phi_i) I_{\lambda, i} \mu_i \, d\mu_i \, d\phi_i}{\int_0^{2\pi} \int_0^1 I_{\lambda, i} \mu_i \, d\mu_i \, d\phi_i}$$

$$(5.3.6)$$

Since $\rho_\lambda(\mu_r, \phi_r)$ is dependent upon the angular distribution of the incident intensity, it is not a surface property. For the special case when the incident radiation is diffuse, the monochromatic hemispherical-angular reflectance is identical to the monochromatic angular-hemispherical reflectance. To prove the equality of $\rho_\lambda(\mu_r, \phi_r)$ and $\rho_\lambda(\mu_i, \phi_i)$ under the condition of constant $I_{\lambda, i}$, it is necessary to use the symmetry of the reflection function as given by

$$\rho_\lambda(\mu_i, \phi_i, \mu_r, \phi_r) = \rho_\lambda(\mu_r, \phi_r, \mu_i, \phi_i) \qquad (5.3.7)$$

and compare Eqs. (5.3.6) (with $I_{\lambda, i}$ independent of incident direction) and (5.3.3). The proof of Eq. (5.3.7) is beyond the scope of this book [see Siegel and Howell (1972)].

The equality of $\rho_\lambda(\mu_i,\phi_i)$ and $\rho_\lambda(\mu_r,\phi_r)$ when $I_{\lambda,i}$ is uniform is of great importance since the measurement of $\rho_\lambda(\mu_r,\phi_r)$ is much easier than $\rho_\lambda(\mu_i,\phi_i)$. By placing a surface in an isothermal cavity, the surface is uniformly irradiated. The energy reflected from the sample through a small observation port in the cavity wall is measured with an instrument which has a small solid angle and which can measure over a small wavelength interval. This is discussed more fully in section 5.5.

Both $\rho_\lambda(\mu_i,\phi_i)$ and $\rho_\lambda(\mu_r,\phi_r)$ can be considered on a total basis by integration over all wavelengths. For the case of the angular-hemispherical reflectance, we have

$$\rho(\mu_i,\phi_i) = \frac{\int_0^\infty q_{\lambda r}\, d\lambda}{\int_0^\infty I_{\lambda,i}\mu_i \Delta\omega_i\, d\lambda} = \frac{1}{\pi I_i}\int_0^\infty \int_0^{2\pi} \int_0^1 \rho_\lambda(\mu_i,\phi_i,\mu_r,\phi_r)I_{\lambda,i}\mu_r$$
$$d\mu_r\, d\phi_r\, d\lambda$$

$$(5.3.8)$$

which, unlike the monochromatic angular-hemispherical reflectance, is not a property.

When a surface element is irradiated from all directions and all the reflected radiation is collected, we characterize this by the monochromatic hemispherical reflectance defined as

$$\rho_\lambda = \frac{q_{\lambda,r}}{q_{\lambda,i}} \qquad (5.3.9)$$

The reflected monochromatic energy $q_{\lambda,r}$ can be expressed in terms of the reflection function and the incident intensity by

$$q_{\lambda,r} = \int_0^{2\pi} \int_0^1 \left[\int_0^{2\pi} \int_0^1 \frac{\rho_\lambda(\mu_r,\phi_r,\mu_i,\phi_i)}{\pi} I_{\lambda,i}\mu_i\, d\mu_i\, d\phi_i \right]\mu_r\, d\mu_r\, d\phi_r$$
$$(5.3.10)$$

The incident energy, expressed in terms of the incident intensity, is

$$q_{\lambda,i} = \int_0^{2\pi} \int_0^1 I_{\lambda,i}\mu_i\, d\mu_i\, d\phi_i \qquad (5.3.11)$$

Division of Eq. (5.3.10) by (5.3.11) yields the monochromatic hemispherical reflectance. For the special case of a diffuse surface (i.e., the reflection function is a constant), the monochromatic hemispherical reflectance is numerically equal to the reflection function and is independent of the spatial distribution of the incident intensity.

The hemispherical reflectance is found by integration of Eqs. (5.3.10) and (5.3.11) over all wavelengths and finding the ratio

$$\rho = \frac{q_r}{q_i} = \frac{\int_0^\infty q_{\lambda,r}\, d\lambda}{\int_0^\infty q_{\lambda,i}\, d\lambda} \tag{5.3.12}$$

The hemispherical reflectance depends on both the angular distribution and wavelength distribution of the incident radiation.

For simple engineering calculations, a special form of the hemispherical reflectance (often the name is shortened to "reflectance")* will be found to be the most useful. The special form is Eq. (5.3.12) with the following assumptions: The reflection function is independent of both direction (diffuse approximation) and wavelength (gray approximation). The diffuse approximation for the hemispherical reflectance has already been discussed and ρ_λ was found to be equal to $\rho_\lambda(\mu_i,\phi_i,\mu_r,\phi_r)$. When the gray approximation is made in addition to the diffuse approximation, the surface reflectance becomes independent of everything except possibly the temperature of the surface, and even this is usually neglected.

5.4 *RELATIONSHIPS AMONG ABSORPTANCE, EMITTANCE, AND REFLECTANCE*

We are now prepared to show that it is necessary to know only one property; namely, the monochromatic angular-hemispherical reflectance, and all absorptance and emittance properties for opaque surfaces can be found.

Let us consider a surface located in an isothermal enclosure maintained at temperature T. The monochromatic intensity in a direction μ,ϕ from an infinitesimal area of the surface consists of emitted and reflected radiation and must be equal

*There are no generally accepted names used in the literature except for the simple "absorptance," "emittance" and "reflectance" which, for clarity, were prefixed with the name hemispherical. In the remainder of this book, the prefix hemispherical will generally be omitted.

to $I_{b\lambda}$ (T).

$$I_{b\lambda} = I_\lambda(\mu,\phi)\Big|_{\text{emitted}} + I_\lambda(\mu,\phi)\Big|_{\text{reflected}} \qquad (5.4.1)$$

The emitted and reflected intensities are

$$I_\lambda(\mu,\phi)\Big|_{\text{emitted}} = \varepsilon_\lambda(\mu,\phi)I_{b\lambda} \qquad (5.4.2)$$

$$I_\lambda(\mu,\phi)\Big|_{\text{reflected}} = I_{b\lambda}\rho_\lambda(\mu,\phi) \qquad (5.4.3)$$

where $\rho_\lambda(\mu,\phi)$ is the monochromatic angular-hemispherical reflectance since the incident intensity is diffuse.

Since $I_{b\lambda}$ can be cancelled from each term, we have

$$\varepsilon_\lambda(\mu,\phi) = 1 - \rho_\lambda(\mu,\phi) \qquad (5.4.4)$$

But from Kirchoff's law,

$$\varepsilon_\lambda(\mu,\phi) = \alpha_\lambda(\mu,\phi) \qquad (5.2.5)$$

or

$$\varepsilon_\lambda(\mu,\phi) = \alpha_\lambda(\mu,\phi) = 1 - \rho_\lambda(\mu,\phi) \qquad (5.4.5)$$

Thus, the monochromatic directional emittance and the monochromatic directional absorptance can both be calculated from knowledge of the monochromatic angular-hemispherical reflectance. All emittance properties [Eqs. (5.1.4), (5.1.6), and (5.1.8)] can then be found once $\rho_\lambda(\mu,\phi)$ is known. The nonproperty absorptances [Eqs. (5.1.2), (5.1.5), and (5.1.7)] can be found if the incident intensity is specified.

5.5   MEASUREMENTS OF SURFACE RADIATION PROPERTIES

In the preceding discussion many radiation surface properties have been defined. Unfortunately, in much of the literature the exact nature of the surface and the type of surface being reported is not clearly specified. This situation requires that caution be exercised.

Many of the reflectance data reported in the literature have been measured by a method devised by Gier et al. (1954).

In this method a cool sample is exposed to blackbody (cavity) radiation from a high-temperature source (a hohlraum) and the monochromatic radiation reflected from the surface is compared to the blackbody radiation from the cavity at that wavelength. The data are thus hemispherical-angular monochromatic reflectances (or angular-hemispherical monochromatic reflectances, since they are equal for diffuse incident radiation). The method is shown schematically in Figure 5.5.1. In many systems the angle between the surface normal and the measured radiation is often fixed at a small value so that measurements can be made at only one angle (approximately normal). In some of the more recent designs the sample can be rotated so that all angles can be measured. With measurements of this type, emittance and absorptance values can be found from Eq. (5.4.5).

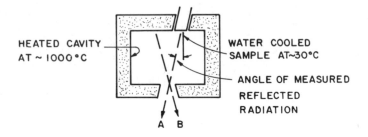

HEATED CAVITY
AT ~ 1000°C

WATER COOLED
SAMPLE AT ~30°C

ANGLE OF MEASURED
REFLECTED
RADIATION

A   B

Figure 5.5.1   Schematic of hohlraum for measurement of mono-chromatic hemispherical-angular reflectance. Radiation B is blackbody radiation from cavity. Radiation A is that blackbody radiation reflected from the cool sample. The ratio A_λ/B_λ, detected by a monochromator, is $\rho_\lambda(\mu,\phi)$.

Table 5.5.1 gives data on surface properties for a few common materials. The data are total hemispherical or total normal emittances at various temperatures and normal solar absorptance at room temperature. Table 5.5.1 was compiled from *Thermophysical Properties of Matter* by Touloukian et al., Vols. 7, 8, and 9 (1970, 1972, and 1972). These three volumes are the most complete reference to radiation properties available today. In addition to total hemispherical and normal emittance, such properties as angular spectral reflectance, angular total reflectance, angular solar absorptance, and others, are given in this extensive compilation.

TABLE 5.5.1 Radiation Properties.

Material		Emittance/Temperature, °K	Absorptance[†]
Aluminum, pure	H*	$\dfrac{0.102}{573}$, $\dfrac{0.130}{773}$, $\dfrac{0.113}{873}$	0.09 - 0.10
Aluminum, Anodized	H	$\dfrac{0.842}{296}$, $\dfrac{0.720}{484}$, $\dfrac{0.669}{574}$	0.12 - 0.16
Aluminum with SiO_2 Coating	H	$\dfrac{0.366}{263}$, $\dfrac{0.384}{293}$, $\dfrac{0.378}{324}$	0.11
Carbon Black in Acrylic Binder	H	$\dfrac{0.83}{278}$	0.94
Chromium	N	$\dfrac{0.290}{722}$, $\dfrac{0.355}{905}$, $\dfrac{0.435}{1072}$	0.415
Copper, polished	H	$\dfrac{0.041}{338}$, $\dfrac{0.036}{463}$, $\dfrac{0.039}{803}$	0.35
Gold	H	$\dfrac{0.025}{275}$, $\dfrac{0.040}{468}$, $\dfrac{0.048}{668}$	0.20 - 0.23
Iron	H	$\dfrac{0.071}{199}$, $\dfrac{0.110}{468}$, $\dfrac{0.175}{668}$	0.44
Lampblack in Epoxy	N	$\dfrac{0.89}{298}$	0.96
Magnesium Oxide	H	$\dfrac{0.73}{380}$, $\dfrac{0.68}{491}$, $\dfrac{0.53}{755}$	0.14
Nickel	H	$\dfrac{0.10}{310}$, $\dfrac{0.10}{468}$, $\dfrac{0.12}{668}$	0.36 - 0.43
Paint			
Parsons Black	H	$\dfrac{0.981}{240}$, $\dfrac{0.981}{462}$	0.98
Acrylic White	H	$\dfrac{0.90}{298}$	0.26
White (ZnO)	H	$\dfrac{0.929}{295}$, $\dfrac{0.926}{478}$, $\dfrac{0.889}{646}$	0.12 - 0.18

*H is total hemispherical emittance; N is total normal emittance.
†Normal Solar Absorptance

5.6 *SELECTIVE SURFACES*

An examination of solar collector energy balances shows the de-
sirability of obtaining surfaces with the combination of high
absorptance for solar radiation and low emittance for long-wave
radiation. This combination of properties is possible to
achieve because there is little overlap in wavelength ranges
between incoming solar energy (outside the earth's atmosphere
98% is at wavelengths less than 3.0 μm) and emitted long-wavelength radiation (less than 1% is at wavelengths less than
3.0 μm for a black surface at 400 °K).
    The concept of a selective surface is shown by considera-
tion of the monochromatic reflectance* as shown in Figure 5.6.1.
This idealized surface is called a semigray surface since it can
be considered gray in the solar spectrum (i.e., at wavelengths
less than about 3.0 μm) and also gray, but with different properties, in the infrared spectrum (i.e., at wavelengths greater
than about 3.0 μm).

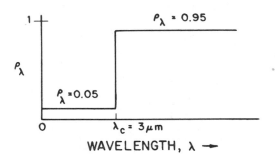

Figure 5.6.1 A hypothetical selective surface.

    For this ideal surface, the monochromatic reflectance ρ_λ
is very low below a critical or cutoff wavelength λ_c, and very
high at wavelengths greater than λ_c. Consequently, the absorptance for solar energy will be very nearly $(1 - \rho_\lambda)$ for $\lambda < 3$ μm.
The value of the emittance will depend on the temperature of the
surface; that is, how much of the radiation is emitted at wavelengths greater than λ_c and how much at wavelengths less than
λ_c. For normal operation of flat-plate solar collectors, the
temperatures will be low enough so that essentially all energy
will be emitted at wavelengths greater than 3 μm.

*Since almost all data is hemispherical, the prefix hemispheri-
cal will not be used.

In practice, the wavelength dependence of ρ_λ does not closely approach the ideal curve of Figure 5.6.1. Examples of ρ_λ versus λ for several real surfaces are shown in Figure 5.6.2. As the real selective surfaces do not have a well-defined critical wavelength λ_c, and uniform properties in the short- and long-wavelength ranges, values of emittance will be more sensitive to surface temperature than those of the ideal semigray surface of Figure 5.6.1. Integration over the emitted spectrum is necessary to estimate the long-wavelength emittance and over the solar spectrum to estimate solar absorptance.*

The potential utility of selective surfaces in solar collectors was inferred by Hottel and Woertz (1942) and more recently discussed by Gier and Dunkle (1958) and by Tabor (1955, 1964,1967). Interest in designing surface with a variety of ρ_λ versus λ characteristics for applications to space vehicles and to solar energy applications resulted in considerable research and compilation of data [e.g., Martin and Bell (1961), Edwards et al. (1960), Schmidt et al. (1964), and others]. Tabor (1967) reviewed selective surfaces and presents several recipes for their preparation. Thus, from 1955 on there have been developments which have applicability to solar collectors, and a number of mechanisms for producing desired combinations of properties have evolved. Among those of interest in solar energy applications are the following:

1. Coatings having high absorptance for solar radiation and high transmittance for long-wave radiaiion can be applied to substrates with low emittance. Thus, in effect, the coating absorbs solar energy, and the substrate is the (poor) emitter of long-wave radiation. Coatings may be homogeneous or have particulate structure; their properties are then either the inherent optical properties of the coating material or of the material properties and the coating structure. Metal oxide coatings on metal substrates which show desirable properties for solar collectors have been developed by Tabor (1967), Hottel and Unger (1959), Kokoropoulos et al. (1959). Finely divided lead sulfide coatings were studied by Williams et al. (1963); the particulate structure of these coatings (as of those of Hottel and Unger) significantly affected the selectivity, and the possibility of a useful selective paint using a vehicle of high transmittance for long-wave radiation was noted.

2. Interference filters can be applied to low emittance substrates. The filters are formed by depositing alternate

*In much of the literature on solar energy the symbol ε is used to indicate emittance in the long-wavelength region and the symbol α is used to indicate solar absorptance.

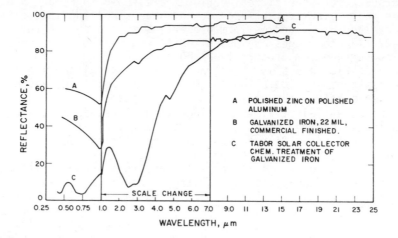

Figure 5.6.2*a* Spectral reflectances for several surfaces.
[From Edwards et al. (1960).]

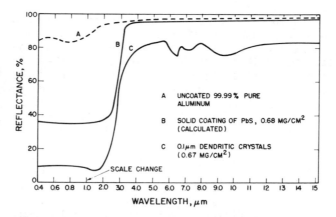

Figure 5.6.2*b* Spectral reflectances for PbS coatings on aluminum substrates. [From Williams et al. (1963).]

layers of metals and dielectrics in quarter wavelength films for
the visible and near-infrared. It has been shown by Martin and
Bell (1960) that three-layer coatings such as SiO_2-Al-SiO_2 on
substrates such as Al should have reflectance less than 0.10 for
solar energy and greater than 0.90 for long-wave radiation.

3. It has been suggested that the surface structure of a
metal normally of high reflectance can be designed to make the
surface a good absorber of short-wave radiation; this is done
by grooving or pitting the surface to create cavities of dimensions near the cutoff wavelength. The surface acts as an assembly of cavity absorbers for short-wavelengths, but for long-wave
radiation radiates as a flat surface. The degree of selectivity
obtainable by this technique has been limited.

4. "Directional selectivity" can be obtained by proper
arrangement of the surface on a large scale. Surfaces of deep
V-grooves, large relative to all wavelengths of radiation concerned, can be arranged so that radiation from near normal directions to the overall surface will be reflected several times
in the grooves, each time absorbing a fraction of the beam.
This multiple absorption gives an increase in the solar absorptance, but at the same time increases the long-wavelength emittance. However, as shown by Hollands (1963), a partially selective surface can have its effective properties substantially improved by proper configuration. For example, a surface having
nominal properties of $\alpha = 0.60$ and $\varepsilon = 0.05$, used in a fixed
optimally oriented flat-plate collector over a year, with 55°
grooves, will have an average effective α of 0.90 and an equivalent ε of 0.10. Figure 5.6.3 from Trombe et al. (1964), illustrates the multiple absorptions obtained for various angles of
incidence of solar radiation on a 30° grooved surface.

The utility of selective surfaces in solar collectors is a
function of two major factors. First, low long-wave emittance
is usually obtained at some sacrifice of high solar absorptance,
and the net effect of selectivity on performance of a collector
(and ultimately on the cost of delivered energy from the collector) must be evaluated for the collector and process in question. This is done by combining energy balance considerations
of Chapter 7 with economic considerations presented in later
chapters.

Second, in practice, solar collectors must be designed to
operate for many years. The surfaces are usually exposed to
oxidizing and corrosive atmospheres, and operate at more or less
elevated temperatures. The data available for α and ε of surfaces are most often available for freshly prepared surfaces.

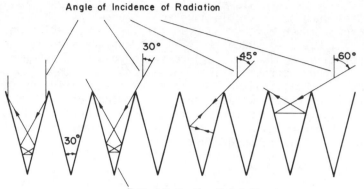

Figure 5.6.3 Absorption of solar radiation by successive reflections on folded metal sheets. [From Trombe et al. (1964)].

Little information is available on radiation properties of surfaces of operating collectors after extended periods of time.
    Table 5.6.1 shows absorptance for solar radiation and emittance for long-wave radiation for temperatures typical of operation of flat-plate collectors. Two surfaces have been applied to commercial solar water heaters; others are experimental.

5.7 "BLACK" SURFACES

The directional absorptance of ordinary blackened surfaces (such as are used for solar collectors) for solar radiation is a function of the angle of incidence of the radiation on the surface. An example of dependence of absorptance on angle of incidence is shown in Figure 5.7.1.

5.8 SPECULARLY REFLECTING SURFACES

Focusing solar collectors require the use of reflecting materials (or possibly refracting materials) to direct the beam component of solar radiation onto a target. This requires surfaces of high specular reflectance for the solar spectrum.
    Specular surfaces are usually metals or metallic coatings on smooth substrates. Opaque substrates must be front-surfaced. Examples are anodized aluminum and rhodium plated on copper. The specular reflectivity of such surfaces is a function of the quality of the substrate and the plating.

TABLE 5.6.1 Properties of Some Selective Surfaces for Solar Energy Application.

Surface	α*	ε*	Reference
"Nickel Black"; containing oxides and sulfides of Ni and Zn, on polished Ni	0.91-0.94	0.11	Tabor et al. (1964)
"Nickel Black" on galvanized iron (experimental)	0.89	0.12	Tabor et al. (1964)
Same process[†] (commercial)		0.16-0.18	Tabor et al. (1964)
"Nickel Black," 2 layers on electroplated Ni on mild steel (α and ε after 6 hr immersion in boiling water)	0.94	0.07	Schmidt (1974)
CuO on Ni; made by electrode position of Cu and subsequent oxidation	0.81	0.17	Kokoropoulos et al. (1959)
Co_3O_4 on silver; by deposition and oxidation	0.90	0.27	Kokoropoulos et al. (1959)
CuO on Al; by spraying dilute $Cu(NO_3)_2$ solution on hot Al plate and baking	0.93	0.11	Hottel and Unger (1959)
"Cu Black"[†] on Cu, by treating Cu with solution of NaOH and $NaClO_2$	0.89	0.17	Close (1962)
Ebanol[†] C on Cu; commercial Cu-blackening treatment giving coatings largely CuO	0.90	0.16	Edwards et al. (1960)
CuO on anodized Al; treat Al with hot $Cu(NO_3)_2$-$KMnO_4$ solution and bake	0.85	0.11	Tabor (1967)
Al_2O_3-Mo-Al_2O_3-Mo-Al_2O_3Mo-Al_2O_3 interference layers on Mo (ε measured at 500 °F)	0.91	0.085	Schmidt et al. (1964)
PbS crystals on Al	0.89	0.20	Williams et al. (1963)

*α = absorptance for solar energy; ε = emittance for long-wave radiation at temperatures typical of flat-plate solar collectors.
[†]Commercial processes.

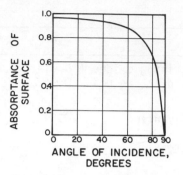

Figure 5.7.1 Directional absorptance of an blackened surface
for artificial sunlight transmitted through glass.

    Specular surfaces can also be applied to transparent sub-
strates, including glass or plastic. If front-surface coatings
are used, the nature of the substrate other than its smoothness
and stability is unimportant. If back-surface coatings are applied, the transparency of the substrate will also be a factor,
as the radiation will pass through the equivalent of twice the
thickness of the substrate and twice through the front surface-
air interface. (See Chapter 6 for discussion of radiation
transmission through partially transparent media.)
    Specular reflectance is, in general, wavelength dependent,
and in principle, monochromatic reflectances should be integrated for the particular spectral distribution of incident
energy. Thus, the monochromatic specular reflectance is defined as

$$\rho_{s,\lambda} = \frac{I_{\lambda,\rho s}}{I_{\lambda,i}} \qquad (5.8.1)$$

where $I_{\lambda,\rho s}$ is the energy specularly reflected at λ, and $I_{\lambda,i}$
is the incident radiation at λ. Then, the specular reflectivity
is

$$\rho_s = \frac{\int_0^\infty \rho_{s,\lambda} I_{\lambda,i} \, d\lambda}{\int_0^\infty I_{\lambda,i} \, d\lambda} \qquad (5.8.2)$$

Typical values of specular reflectance of surfaces of solar radiation are shown in Table 5.8.1.

TABLE 5.8.1 Normal Specular Solar Reflectances of Surfaces for Beam Radiation.

Surface	ρ
Electroplated silver, new	0.96
High-purity Al, new clean	0.91
Sputtered aluminum optical reflector	0.89
Brytal processed aluminum, high purity	0.89
Back-silvered water white plate glass, new, clean	0.88
Al, SiO coating, clean	0.87
Aluminum foil, 99.5% pure	0.86
Back-aluminized 3M acrylic, new	0.86
Commercial Alzac process aluminum	0.85
Back-aluminized 3M acrylic, new	0.85*
Aluminized Type C Mylar (from Mylar Side)	0.76

*Exposed to equivalent of 1 yr solar radiation.

The table includes data on front-surface and second-surface reflectors. The aluminized acrylic film is one of a number of aluminized polymeric films that have been evaluated for durability in weather and it appears to be the best of those reported by Honeywell (1973). Many other such materials have short life-times (on the order of weeks or months) under practical operating conditions.

The maintenance of high specular reflectance presents practical problems. Front-surface reflectors are subject to degradation by oxidation, abrasion, dirt, and so on. Back-surfaced reflectors may lose reflectance because of dirt or degradation of the overlying transparent medium.

Front-surface reflectors may be covered by thin layers of protective materials to increase their durability. For example, anodized aluminum is coated with a thin stable layer of aluminum oxide deposited by electrochemical means and silicon monoxide has been applied to front-surfaced aluminum films. In general, each coating reduces the initial value of specular reflectance, but may result in maintenance of more satisfactory levels of reflectance over long periods of time.

REFERENCES

Close, D. J., Report E.D.7, Engr. Sect., Commonwealth Scientific and Industrial Research Organization, Melbourne, Australia (1962). "Flat Plate Solar Absorbers: The Production and Testing of a Selective Surface for Copper Absorber Plates."

Edwards, D. K., Nelson, K. E., Roddick, R. D., and Gier, J. T., Report No. 60-93, Dept. of Engineering, Univ. of California (October 1960). "Basic Studies on the Use and Control of Solar Energy."

Gier, J. T., Dunkle, R. V., and Bevans, J. T., J. Opt. Soc. Am., *44*, 558 (1954). "Measurement of Absolute Spectral Reflectivity from 1.0 to 15 Microns."

Gier, J. T. and Dunkle, R. V. <u>Transactions of the Conference on the Use of Solar Energy</u>, *2*, Part I, 41, U. of Arizona Press, 1958. "Selective Spectral Characteristics as an Important Factor in the Efficiency of Solar Collectors."

Hollands, K. G. T., Solar Energy, *7*, 108 (1963). "Directional Selectivity, Emittance and Absorptance Properties of Vee Corrugated Specular Surfaces."

Hottel, H. C. and Unger, T. A., Solar Energy, *3*, 10, No. 3 (1959). "The Properties of a Copper Oxide-Aluminum Selective Black Surface Absorber of Solar Energy."

Hottel, H. C., Woertz, B. B., Trans. ASME *14*, 91 (1942). "Performance of Flat-Plate Solar-Heat Exchangers."

Kokoropoulos, P., Salem, E., and Daniels, F., Solar Energy, *3*, 19, No. 4 (1959). "Selective Radiation Coatings-Preparation and High Temperature Stability."

Martin, D. C. and Bell, R., Proceedings of the Conference on Coatings for the Aerospace Environment, WADD-TR-60-TB, Wright Air Development Division, Dayton, Ohio (November 1960). "The Use of Optical Interference to Obtain Selective Energy Absorption."

Schmidt, R. N., Park, K. C., and Janssen, E., Tech. Doc. Report No. ML-TDR-64-250 from Honeywell Research Center to Air Force Materials Laboratory (September 1964). "High Temperature Solar Absorber Coatings, Part II."

Schmidt, R. N., Honeywell Corp., private communication, 1974.

Siegel, R., Howell, J. R., Thermal Radiation Heat Transfer, New York, McGraw-Hill, 1972.

Tabor, H., Bull. Res. Coun. Israel, 5A, No. 2, 119 (January 1956). "Selective Radiation."

Tabor, H., Low Temperature Engineering Applications of Solar Energy, New York, ASHRAE, 1967. "Selective Surfaces for Solar Collectors."

Tabor, H., Harris, J., Weinberger, H., and Doron, B., Proceedings UN Conference on New Sources of Energy, 4, 618, (1964). "Further Studies on Selective Black Coatings."

Touloukian, Y. S., et al. Thermophysical Properties of Matter Plenum Data Corporation, Vol. 7: "Thermal Radiative Properties-Metallic Elements and Alloys," (1970); Vol. 8: "Thermal Radiative Properties-Nonmetallic Solids," (1972); Vol. 9: "Thermal Radiative Properties-Coatings," (1972).

Trombe, F., Foex, M., and LePhat Vinh, M., Proceedings of the UN Conference on New Sources of Energy, 4, 625, 638 (1964). "Research on Selective Surfaces for Air Conditioning Dwellings."

University of Minnesota and Honeywell Corp. Progress Report No. 2 to the National Science Foundation, NSF/RANN/SE/GI-34871/PR/73/2 (1973). "Research Applied to Solar-Thermal Power Systems."

Williams, D. A., Lappin, T. A., and Duffie, J. A., Trans. ASME, J. Engr. Power, 85A, 213 (1963). "Selective Radiation Properties of Particulate Coatings."

6. TRANSMISSION OF RADIATION THROUGH PARTIALLY TRANSPARENT MEDIA

For opaque surfaces, the sum of absorptance and reflectance must be unity. If the surface is transparent to the incident radiation to any degree, the sum of absorptance, reflectance, and transmittance must be unity (i.e., the incident radiation is absorbed, reflected, or transmitted). This relationship holds for surfaces, or layers of a material of finite thickness. Transmittance, like reflectance and absorptance, is a function of wavelength, angle of incidence of the incoming radiation, the refractive index, n, and the extinction coefficient, K, of the material. Strictly speaking, n and K are functions of the wavelength of the radiation, but for most solar energy applications they can be assumed to be independent of λ. Useful reviews of important considerations of transmission of solar radiation have been presented by Dietz (1954,1963).

6.1 REFLECTION AT INTERFACES

Fresnel has derived a relation for the reflection of nonpolarized radiation on passing from a medium 1 with refractive index n_1 to medium 2 with refractive index n_2

$$\frac{I_r}{I_0} = \rho = \frac{1}{2}\left[\frac{\sin^2(\theta_2 - \theta_1)}{\sin^2(\theta_2 + \theta_1)} + \frac{\tan^2(\theta_2 - \theta_1)}{\tan^2(\theta_2 + \theta_1)}\right] \qquad (6.1.1)$$

where θ_1 and θ_2 are the angles of incidence and refraction, as shown in Figure 6.1.1. In this expression the two terms in the square brackets represent the reflection for each of the two components of polarization. Equation (6.1.1), then, gives the reflection of nonpolarized radiation as the average of the two components. The angles θ_1 and θ_2 are related to the indices of refraction by Snell's law

$$\frac{n_1}{n_2} = \frac{\sin \theta_2}{\sin \theta_1} \qquad (6.1.2)$$

Thus if the angle of incidence and refractive indices are known, Eqs. (6.1.1) and (6.1.2) are sufficient to calculate the reflectance of the single interface.

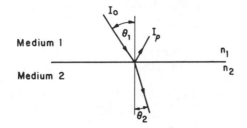

Medium 1

Medium 2

Figure 6.1.1   Angles of incidence and refraction in media having refractive indices n_1 and n_2.

For radiation at normal incidence, both θ_1 and θ_2 are 0, and Eqs. (6.1.1) and (6.1.2) can be combined to yield

$$\rho = \frac{I_r}{I_0} = \left[\frac{(n_1 - n_2)}{(n_1 + n_2)}\right]^2 \qquad (6.1.3)$$

If one medium is air (i.e., a refractive index of unity) Eq. (6.1.3) becomes

$$\rho = \frac{I_r}{I_0} = \left[\frac{(n - 1)}{(n + 1)}\right]^2 \qquad (6.1.4)$$

Example 6.1.1   Calculate the reflectance of one surface of glass at normal incidence and at 60°. The average index of refraction of glass for the solar spectrum is 1.526.

At normal incidence Eq. (6.1.4) can be used:

$$\rho(0) = \left[\frac{0.526}{2.526}\right]^2 = 0.0434$$

At an incident angle of 60°, Eq. (6.1.2) gives the refraction angle, θ_2.

$$\theta_2 = \arcsin\left[\frac{\sin 60}{1.526}\right] = 34.57$$

From Eq. (6.1.1) the reflection is

$$\rho(60) = \frac{1}{2}\left[\frac{\sin^2(-25.42)}{\sin^2(94.57)} + \frac{\tan^2(-25.42)}{\tan^2(94.57)}\right]$$

$$= \frac{1}{2}[0.185 + 0.001] = 0.093$$

Cover materials used in solar applications require the transmission of radiation through a slab or film of material, and there are thus two interfaces per cover to cause reflection loss. In this situation, the depletion of the beam at the second surface is the same as that at the first, for each component of polarization, assuming the cover interfaces are with air on both sides.

Neglecting absorption in the slab shown in Figure 6.1.2, $(1 - \rho)$ of the incident beam reaches the second interface. Of this, $(1 - \rho)^2$ passes through the interface and $\rho(1 - \rho)$ is reflected back to the first, and so on. Summing up the resulting terms, the transmittance for a single cover neglecting absorption is

$$\tau_{r,1} = (1 - \rho)^2 \sum_{n=0}^{\infty} \rho^{2n} = \frac{(1 - \rho)^2}{(1 - \rho^2)} = \frac{1 - \rho}{1 + \rho} \qquad (6.1.5)$$

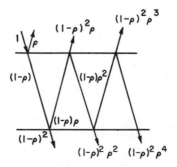

Figure 6.1.2 Transmission through one cover.

For a system of n covers, all of the same material, a similar analysis yields

$$\tau_{r,n} = \frac{(1 - \rho)}{1 + (2n - 1)\rho} \qquad (6.1.6)$$

This relationship holds for each of the two components of polarization. The transmittance for initially unpolarized light is found by taking the average transmittance of the two components.*

Example 6.1.2 Calculate the transmittance due to reflection of two covers of glass at zero incidence and at 60°.

At zero incidence, the reflectance of one interface from Eq. 6.1.1 is 0.0434. With Eq. (6.1.6), the transmittance due to reflection only, is

$Eq\ 6.1.3$

$$\tau_r(0) = \frac{1 - 0.0434}{1 + 3(0.0434)} = 0.85 \qquad n = 2$$

At a 60° incidence, the reflectances of one interface for each component of polarization are 0.185 and 0.001. With Eq. (6.1.6) used for each component of polarization, the transmittance is found from

$$\tau_r(60) = \frac{1}{2}\left[\frac{1 - 0.185}{1 + 3(0.185)} + \frac{1 - 0.001}{1 + 3(0.001)}\right] = 0.76$$

The solar transmittance (due to reflection) of glass, with an average refractive index of 1.526 in the solar spectrum, has been calculated for all incidence angles in the same manner as illustrated in Examples 6.1.1 and 6.1.2. The results for from one to four glass covers are given in Figure 6.1.3. This figure is a recalculation of the results presented by Hottel and Woertz (1942).

6.2 ABSORPTION OF RADIATION IN PARTIALLY TRANSPARENT MEDIA

The absorption of radiation in a partially transparent medium is described by Bouger's law, which is based on the assumption that the absorbed radiation is proportional to the local intensity in the medium and the distance the radiation travels in the medium, x:

$$dI = IK\ dx \qquad (6.2.1)$$

*For angles less than about 40°, the transmittance of a cover system can be estimated using the average reflectance, as calculated from Eq. (6.1.1), in Eq. (6.1.6).

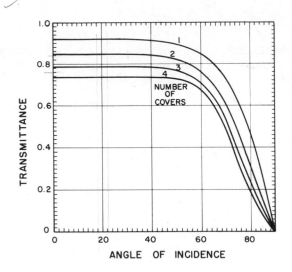

Figure 6.1.3 Transmittance neglecting absorption (i.e., due to reflection only) of 1, 2, 3 and 4 covers having an index of refraction of 1.526.

where K is the extinction coefficient and is assumed to be a constant in the solar spectrum. Integrating this between limits of 0 and L,

$$\tau_a = \frac{I_L}{I_0} = e^{-KL} \qquad (6.2.2)$$

For glass, the value of K varies from about 0.04/cm for "water white" glass to about 0.32/cm for poor (greenish cast of edge) glass. Note that τ_a is the transmittance considering only absorption, and that L is the actual path of the radiation through the medium.

To obtain the transmittance allowing for both reflection and absorption, it is only necessary to multiply the two transmittances together.

$$\tau = \tau_r \tau_a \qquad (6.2.3)$$

This is a satisfactory relationship provided the product KL is small (i.e., τ_a is not too far from unity).* This condition is

*See problem 6-4.

always met in solar collectors at angles of practical interest.

Example 6.2.1 Calculate the total solar transmittance at incidence angles of zero and 60° for two glass covers each 0.23 cm thick. The extinction coefficient of the glass is 0.161/cm and the refractive index is 1.526.

For one sheet at normal incidence, the KL product is

$$KL = 0.23 \times 0.161 = 0.0370$$

The transmittance due to absorption is then

$$\tau_a(0) = e^{-2(0.0370)} = 0.93$$

The transmittance due to reflection, from Example 6.1.2, is 0.85. The total transmittance is then found from Eq. (6.2.3):

$$\tau(0) = \tau_r(0)\tau_a(0) = 0.85(0.93) = 0.79$$

From example 6.1.1, when $\theta_1 = 60°$, $\theta_2 = 34.57$, and

$$\tau_a(60) = e^{-2(0.0370)/\cos 34.57} = 0.91$$

and the total transmittance (with $\tau_r = 0.76$ from Example 6.1.2) becomes

$$\tau(60) = \tau_r(60)\tau_a(60) = 0.76(0.91) = 0.69$$

Figure 6.2.1 gives curves of transmittance as a function of angle of incidence for systems of one to four covers, for three different kinds of glass. These curves were calculated from Eq. (6.2.3) and have been checked by experiments [Hottel and Woertz (1942)].

6.3 TRANSMITTANCE-ABSORPTANCE PRODUCT

To use the analysis of the next chapter, it is necessary to evaluate the transmittance-absorptance product $(\tau\alpha)$. Of the radiation passing through the cover system and striking the plate, some is reflected back to the cover system. However, all this radiation is not lost since some is reflected back to the plate.

The situation is illustrated in Figure 6.3.1 where τ is

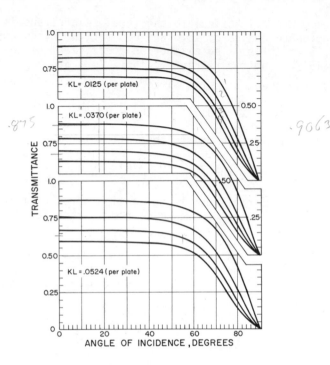

Figure 6.2.1 Transmittance (considering absorption and reflection) of 1, 2, 3 and 4 covers for three types of glass.

the transmittance of the cover system at the desired angle as
calculated from Eq. (6.2.3) and α is the angular absorptance
of the absorber plate. Of the incident energy, $\tau\alpha$ is absorbed
by the absorber plate and $(1 - \alpha)\tau$ is reflected back to the
cover system. The reflection from the absorber plate is probably more diffuse than specular so that the fraction $(1 - \alpha)\tau$
that strikes the cover plate is diffuse radiation and $(1 - \alpha)\tau\rho_d$
is reflected back to the absorber plate. The quantity ρ_d refers to reflection of the cover plate for incident-diffuse radiation that may be partially polarized due to reflections as
it is passed through the cover system. The multiple reflection
of diffuse radiation continues so that the energy ultimately
absorbed is*

*Note that the absorptance, α, of the absorber plate for the re-
flected radiation should probably be the absorptance for diffuse
radiation. The resulting error is small.

$$(\tau\alpha) = \tau\alpha \sum_{n=0}^{\infty} [(1 - \alpha)\rho_d]^n = \frac{\tau\alpha}{1 - (1 - \alpha)\rho_d} \qquad (6.3.1)$$

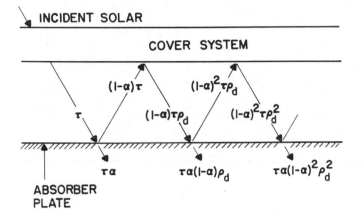

Figure 6.3.1   Absorption of solar radiation by absorber plate.

The diffuse reflectance, ρ_d, can be estimated by using the specular reflection of the cover system at an incidence angle of 60°. From Figure 6.1.3 for one, two, three, and four glass cover systems, ρ_d is approximately 0.16, 0.24, 0.29, and 0.32, respectively.

Example 6.3.1   For a three-cover collector using glass with KL = 0.0524 per plate and an absorber plate with α = 0.94 (independent of direction), find the transmittance-absorptance product at an angle of 55°.

From Figure 6.2.1 at 55°, τ = 0.57. From Figure 6.1.3 but at 60° ρ_d = (1. - 0.71) = 0.29. Therefore, from Eq. (6.3.1),

$$(\tau\alpha) = \frac{0.94 \times 0.57}{1 - 0.06 \times 0.29} = 0.55$$

6.4   *SPECTRAL DEPENDENCE OF TRANSMITTANCE*

Most transparent media transmit selectively; that is, transmittance is a function of wavelength of the incident radiation. Glass, the material most commonly used as a cover material in

solar collectors, may absorb little of the solar energy spectrum
if its Fe_2O_3 content is low; if Fe_2O_3 is high, it will absorb in
the infrared portion of the solar spectrum. The transmittance
of several glasses of varying iron content is shown in Figure
6.4.1. These show clearly that "water white," low-iron glass
has the best transmission; glasses with high Fe_2O_3 have a greenish appearance and are relatively poor transmitters. Note that
the transmission is not a strong function of wavelength in the
solar spectrum except for the "heat absorbing" glass. The
curves also show that glass becomes substantially opaque at
wavelengths longer than about 3 μm so that glass can be considered as opaque to long-wave radiation.

Plastic films may be used as cover materials for solar
collectors. In general, they will have transmittances which
are more wavelength dependent than glass, and it may be necessary to obtain their transmittance for solar energy by evaluating Eq. (6.2.3) for monochromatic radiation and then integrating
over the solar spectrum. The total angular transmittance then
becomes

$$\tau(\mu,\phi) = \frac{\int_0^\infty \tau_\lambda I_{\lambda,i}(\mu,\phi)\ d\lambda}{\int_0^\infty I_{\lambda,i}(\mu,\phi)\ d\lambda} \qquad (6.4.1)$$

If the absorptance of solar radiation by an absorber plate
is independent of wavelength, then Eq. (6.3.1) can be used to
find the transmittance-absorptance product with the transmittance
as calculated from Eq. (6.4.1). However, if both the solar
transmittance and solar absorptance are functions of wavelength,
the fraction absorbed by an absorber plate is given by

$$\tau\alpha = \frac{\int_0^\infty \tau_\lambda \alpha_\lambda I_{\lambda,i}(\mu,\phi)\ d\lambda}{\int_0^\infty I_{\lambda,i}(\mu,\phi)\ d\lambda} \qquad (6.4.2)$$

To account accurately for multiple reflections in a manner analogous to Eq. (6.3.1), it would be necessary to evaluate the
spectral distribution of each reflection and integrate overall
wavelengths. It is unlikely that such a calculation would ever
be necessary for solar collector systems since the error involved by directly using Eq. (6.4.2) with (6.3.1) would be small
if α is near unity.

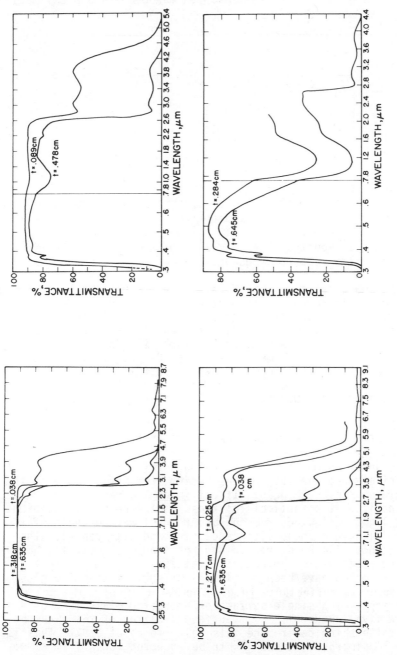

Figure 6.4.1 Spectral transmittance of glass. Upper left, glass containing 0.02 Fe_2O_3; Upper right, 0.50 Fe_2O_3. Lower left, 0.10 Fe_2O_3. Lower right, 0.15 Fe_2O_3. [From Dietz (1954)].

117

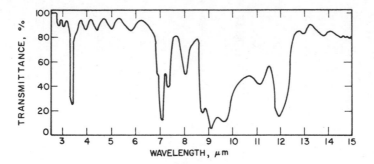

Figure 6.4.2   Spectral transmittance of a polyvinyl fluoride
"Tedlar" film.   (Courtesy duPont).

For most plastics, the transmittance will also be significant in the infrared spectrum at λ > 3 μm. Figure 6.4.2 shows
a transmittance curve for a polyvinyl fluoride ("Tedlar") film
for wavelengths longer than 2.5 μm. Whillier (1963) calculated
the transmittance of a similar "Tedlar" film for radiation from
blackbody sources at temperatures from 0 °C to 200 °C. He
found that transmittance was 0.32 for radiation from the blackbody source at 0 °C, 0.29 for the source at 100 °C, and 0.32
for the source at 200 °C.

6.5 *EFFECTS OF SURFACE LAYERS ON TRANSMITTANCE*

By the addition of thin films having a refractive index between
that of air and the transparent medium, the reflectance of the
interfaces can be changed. If a film of low refractive index
is deposited at an optical thickness of λ/4 onto a transparent
slab, light reflected from the upper and lower surfaces of the
film will have a phase difference of π and will cancel. The
reflectance will be decreased and the transmittance will be increased relative to the uncoated material.

Processes have been developed for surface treating glass
to reduce its reflectance [e.g., Thomsen (1951)]. The solar
reflectance of a single surface of untreated glass is 0.0434
(refractive index = 1.526). Surface treating of the glass by
chemical etching can reduce this value to about 0.02. This is
a greater improvement than is to be expected for a film of optimum thickness. Since the radiation covers a wavelength range
and both constructive and destructive interference occurs, the

surface-treating process must produce a gradation of refractive index to account for the change. An increase of transmittance from 0.92 to 0.96 can make a very significant improvement in the thermal performance of flat-plate collectors. However, the cost of large-scale surface treatment of glass is unknown at this time.

REFERENCES

Dietz, A. G. H., Space Heating with Solar Energy, R. W. Hamilton, ed., Massachusetts Institute of Technology, 1954. "Diathermanous Materials and Properties of Surfaces."

Dietz, A. G. H., Introduction to the Utilization of Solar Energy, Zarem, A. M. and Erway, D. D., Eds., New York, McGraw-Hill, 1963. "Diathermanous Materials and Properties of Sur-Faces."

Hottel, H. C. and Woertz, B. B., Trans. ASME, 64, 91 (1942). "The Performance of Flat-Plate Solar-Heat Collectors."

Thomsen, S. M., RCA Review, 12, 143 (1951). "Low-Reflection Films Produced on Glass in a Liquid Fluosilicic Acid Bath."

Whillier, A., Solar Energy, 7, 148 (1963). "Plastic Covers for Solar Collectors."

7. FLAT-PLATE COLLECTORS

The solar collector is the essential item of equipment which transforms solar radiant energy to some other useful energy form. A solar collector differs in several respects from more conventional heat exchangers. The latter usually accomplish a fluid-to-fluid exchange with high heat transfer rates and with radiation as an unimportant factor. In the solar collector, energy transfer is from a distant source of radiant energy to a fluid. Without optical concentration, the flux of incident radiation is, at best, about 1100 W/m^2 and is variable. The wavelength range is from 0.3 to 3.0 μm, which is considerably shorter than that of the emitted radiation from most energy-absorbing surfaces. Thus, the analysis of solar collectors presents unique problems of low and variable energy fluxes and the relatively large importance of radiation.

Solar collectors may be used with or without radiation concentration. For flat-plate collectors the area absorbing solar radiation is the same as the area intercepting solar radiation. Focusing collectors usually have concave reflectors to concentrate the radiation falling on the total area of the reflector onto a heat exchanger of smaller surface area, thereby increasing the energy flux. In this chapter we discuss only flat-plate collectors; focusing collectors are treated in the following chapter.

Flat-plate collectors can be designed for applications requiring energy delivery at moderate temperatures, up to perhaps 100 °C above ambient temperature. They have the advantages of using both beam and diffuse solar radiation, not requiring orientation toward the sun, and requiring little maintenance. They are mechanically simpler than the concentrating reflectors, absorbing surfaces, and orientation devices of focusing collectors. The principle present applications of these units are in solar water heating systems, while potential uses include building heating and air conditioning.

The importance of flat-plate collectors in thermal processes is such that their thermal performance is treated here in considerable detail. This is done to develop, for the user of this book, an understanding of how the component functions. In many practical cases of design calculations, the formulations

of collector performance are reduced to relatively simple form.
For the user who is already familiar with the principles of collector operation, we draw attention to section 7.13, Short-Term
Collector Performance, which summarizes the detailed discussions
of this chapter.

Costs of collectors are not considered in this chapter.
The cost of energy delivered from a solar collector will depend
on the system thermal performance, on first costs and on component lifetimes. These aspects are considered in later chapters for specific applications.

7.1 GENERAL DESCRIPTION OF FLAT-PLATE COLLECTORS

The important parts of a typical flat-plate solar collector, as
shown in Figure 7.1.1, are: the "black" solar energy-absorbing
surface, with means for transferring the absorbed energy to a
fluid; envelopes transparent to solar radiation over the solar
absorber surface which reduce convection and radiation losses
to the atmosphere; and back insulation to reduce conduction
losses as the geometry of the system permits. Although Figure
7.1.1 depicts a water heater, and most of the analysis of this
chapter is concerned with this geometry, air heaters are funda-
mentally the same and are discussed in section 7.12.

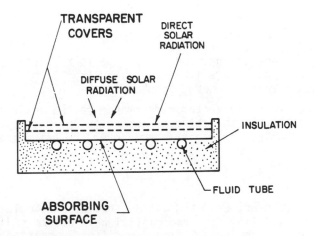

Figure 7.1.1 Basic flat-plate solar collector.

Flat-plate collectors are usually, although not always, mounted in a stationary position (e.g., as an integral part of a wall or roof structure in solar house heating) with an orientation optimized for the particular location in question for the time of year in which the solar device is intended to operate. In their most common forms, they are air or water heaters or low-pressure steam generators.

7.2 THE BASIC FLAT-PLATE ENERGY BALANCE EQUATION

The performance of a solar collector is described by an energy balance that indicates the distribution of incident solar energy into useful energy gain and various losses. The energy balance on the whole collector can be written as

$$A_c\{[HR(\tau\alpha)]_b + [HR(\tau\alpha)]_d\} = Q_u + Q_L + Q_s \qquad (7.2.1)$$

where H = rate of incidence of beam or diffuse radiation on a unit area of surface of any orientation;

R = factor to convert beam or diffuse radiation, to that on the plane of the collector,*

$(\tau\alpha)$ = transmittance-absorptance product of cover system for beam or diffuse radiation;

A_c = collector area;

Q_u = rate of useful heat transfer to a working fluid in the solar exchanger;

Q_L = rate of energy losses from the collector to the surroundings by reradiation, convection, and by conduction through supports for the absorber plate, and so on. The losses due to reflection from the covers are included in the $(\tau\alpha)$ term above; and

Q_s = rate of energy storage in the collector.

A measure of collector performance is the collection efficiency, defined as the ratio of the useful gain over any time period to the incident solar energy over the same time period.

$$\eta = \int \frac{Q_u}{A_c}\, d\tau \Big/ \int HR\, d\tau \qquad (7.2.2)$$

*Strictly speaking, beam and diffuse radiation should be treated separately. $(\tau\alpha)$ for beam radiation is determined from the actual angle of incidence; $(\tau\alpha)$ for diffuse radiation may be taken as that for beam radiation at an incidence angle of 60°. Throughout most of this chapter, the symbol HR is used to represent the sum of $H_b R_b$ and $H_d R_d$.

In the subsequent sections of this chapter, energy balances like
Eq. (7.2.1) will be applied to various types of collectors with
the objectives of calculating the rate of useful energy gain and
the collection efficiency.

It should be pointed out that the design of a solar energy
system is concerned with obtaining minimum energy cost. Thus,
it may be desirable to design a collector with an efficiency
lower than is technologically possible if the cost is significantly reduced. In any event, it is necessary to predict the
performance of a collector and that is the basic aim of this
chapter.

7.3 GENERAL CHARACTERISTICS OF FLAT-PLATE SOLAR COLLECTORS

The detailed analysis of a solar collector is a very complicated
problem. Fortunately, a relatively simple analysis will yield
very useful results. These results will show the important
variables, how they are related, and how they affect the performance of a solar collector. To illustrate these basic principles, a simple configuration, as shown in Figure 7.3.1, will
first be examined. The analysis presented follows the basic
derivation by Hottel and Whillier (1958), Bliss (1959), Whillier
(1953,1967).

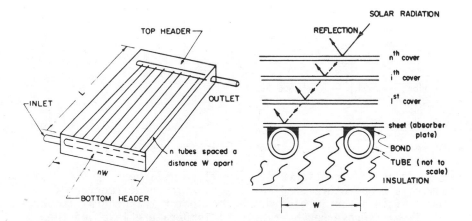

Figure 7.3.1 Sheet and tube solar collector.

To appreciate more fully the mathematics that will be

presented, it is desirable to have an understanding of the temperature gradients that exist in a solar collector constructed
as shown in Figure 7.3.1. Figure 7.3.2 shows the region between
two tubes. Some of the solar energy absorbed by the plate must
be conducted along the plate to the region of the tubes. Thus,
the temperature midway between the tubes will be higher than the
temperature in the vicinity of the tubes. The temperature above
the tubes will be nearly uniform because of the presence of the
tube and weld metal.

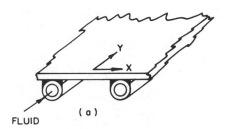

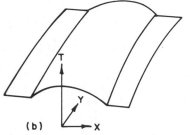

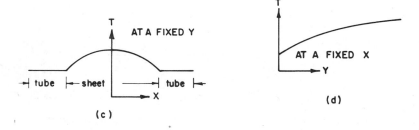

Figure 7.3.2 Temperature distribution of absorber plate.

The energy transferred to the fluid will heat up the fluid
causing a temperature gradient to exist in the direction of
flow. Since in any region of the collector the general temperature level is governed by the local temperature level of the
fluid, a situation such as shown in Figure 7.3.2b is expected.
At any location y, the general temperature distribution in the
x direction is as shown in Figure 7.3.2c, and at any location x,

the temperature distribution in the y direction will look like Figure 7.3.2d.

To model the situation shown in Figure 7.3.2, a number of simplifying assumptions can be made to lay the foundations without obscuring the basic physical situation. These important assumptions are:

1. Performance is steady-state.
2. Construction is of sheet and tube type.
3. The headers cover a small area of collector and can be neglected.
4. The headers provide uniform flow to tubes.
5. There is no absorption of solar energy by covers insofar as it affects losses from the collector.
6. There is one-dimensional heat flow through covers.
7. There is a negligible temperature drop through a cover.
8. There is one-dimensional heat flow through back-insulation.
9. The sky can be considered as a blackbody for long-wavelength radiation at an equivalent sky temperature.
10. Temperature gradients around tubes can be neglected.
11. The temperature gradients in the direction of flow and between the tubes can be treated independently.
12. Properties are independent of temperature.
13. Loss through front and back are to the same ambient temperature.
14. Dust and dirt on the collector are negligible.
15. Shading of the collector absorbing plate is negligible.

In later sections of this chapter many of the above assumptions will be relaxed.

7.4 COLLECTOR OVERALL HEAT TRANSFER COEFFICIENT

It is desirable to develop the concept of an overall loss coefficient for a solar collector to simplify the mathematics. Let us consider the thermal network for a three-cover system shown in Figure 7.4.1. At some typical location on the plate where the temperature is T_p, solar energy of amount S is absorbed by the plate; S is equal to $[HR(\tau\alpha)]_{beam}$ plus $[HR(\tau\alpha)]_{diffuse}$. This absorbed energy S is distributed to losses through the top, bottom, and edges and to useful energy gain. The purpose of this section is to convert the thermal network of Figure 7.4.1 to the thermal network of Figure 7.4.2.

The energy loss through the bottom of the collector is represented by two series resistors, R_1 and R_2. R_1 represents the

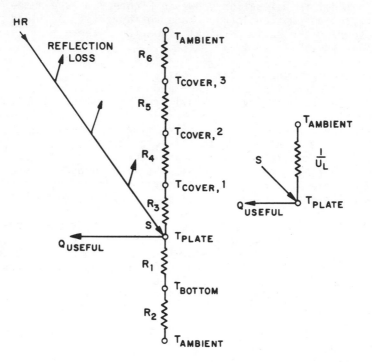

Figure 7.4.1 Thermal network for flat-plate solar collector.

Figure 7.4.2 Equivalent thermal network for flat-plate solar collector.

resistance to heat flow through the insulation and R_2 represents the convection and radiation resistance to the environment. The magnitudes of R_1 and R_2 are such that it is usually possible to assume R_2 is zero and all resistance to heat flow is due to the insulation. Thus, the back loss coefficient, U_b, is approximately*

$$U_b = \frac{1}{R_1} = \frac{k}{L} \qquad\qquad (7.4.1)$$

where k and L are the insulation thermal conductivity and thickness respectively.

*The back losses may not be to the same temperature as the top losses.

For most collectors the evaluation of edge losses is very complicated. However, in a well-designed system the edge loss should be small so that it is not necessary to predict it with great accuracy. Tabor (1958) recommends edge insulation of about the same thickness as bottom insulation. The edge losses are then estimated by assuming one-dimensional sideways heat flow around the perimeter of the collector system.

The loss coefficient for the top surface is the result of convection and radiation between parallel plates. The energy transfer between the plate at T_p and the first cover glass at $T_{cover,1}$ is exactly the same as between any other two adjacent glass plates and is also equal to the energy lost to the surroundings from the top cover glass. (This is not true at any instant of time if energy storage in the glass is considered or if the glass absorbs solar energy.) The loss through the top per unit area is then

$$q_{loss,top} = h_{p-c1}(T_{plate} - T_{cover,1})$$

$$+ \frac{\sigma(T_{plate}^4 - T_{cover,1}^4)}{(1/\varepsilon_{plate}) + (1/\varepsilon_{glass}) - 1}$$

$h_r \rightarrow page\ 77$

$$(7.4.2)$$

where h_{p-c1} is the heat transfer coefficient between two inclined parallel plates from Chapter 4. If the radiation term is linearized, the radiation heat transfer coefficient can be used and the heat loss becomes

$$q_{loss,top} = (h_{p-c1} + h_{r1})(T_{plate} - T_{cover,1}) \quad (7.4.3)$$

where $h_{r1} = \sigma(T_{plate} + T_{cover,1})(T_{plate}^2 + T_{cover,1}^2)/[(1/\varepsilon_{plate}) + (1/\varepsilon_{glass}) - 1]$. The resistance, R_3, can then be expressed as

$$R_3 = \frac{1}{h_{p-c1} + h_{r1}} \quad (7.4.4)$$

A similar expression can be written for each resistance between the glass plates. To a reasonable degree of approximation, the resistances R_4 and R_5 are equal. These resistances are not generally equal to R_3 since R_3 contains the plate emittance which is not equal to the glass emittance. The plate may be a selective surface such that it has a high absorptance for solar radiation and a low emittance for long-wavelength radiation. In general, we can have as many cover plates as desired but the practical limit seems to be three with most systems using one or two.

The last resistance from the top cover to the surroundings has the same form as Eq. (7.4.4), but the convection heat transfer coefficient is for wind blowing over the collector. Approximate values for this coefficient are given by Eq. (4.14.4). The radiation resistance from the top cover accounts for radiation exchange with the sky at T_{sky}. For simplicity, we will reference this resistance to the air temperature so that the radiation conductance can be written as

$$h_{r6} = \varepsilon_{glass}\sigma(T_{cover,n} + T_{sky})(T^2_{cover,n} + T^2_{sky})$$

$$\left(\frac{T_{cover,n} - T_{sky}}{T_{cover,n} - T_{ambient}}\right)$$

$$(7.4.5)$$

where $T_{cover,n}$ is the temperature of the top cover. The resistance to the surroundings is then given by

$$R_6 = \frac{1}{h_w + h_{r6}} \qquad (7.4.6)$$

For this three-cover system the top loss coefficient from the collector plate to the ambient is

$$U_t = \frac{1}{R_3 + R_4 + R_5 + R_6} \qquad (7.4.7)$$

Example 7.4.1 Calculate the top loss coefficient for a single glass cover with the following specifications:

Plate to cover spacing	2.5 cm
Plate emittance	0.95
Ambient air and sky temperature	10 °C
Wind speed	5.0 m/s
Back insulation thickness	5 cm
Insulation conductivity	0.045 W/mC
Mean plate temperature	65 °C
Collector tilt	45°

The procedure for solving for the top loss coefficient is necessarily an iterative process. For this single glass cover system, Eq. (7.4.7) becomes

$$U_t = \left(\frac{1}{h_{p-c} + h_{r,p-c}} + \frac{1}{h_w + h_{r,c-s}} \right)^{-1}$$

The convection coefficient between the plate and the cover, h_{p-c}, can be found using Eq. (4.11.4a). The wind convection coefficient, h_w, can be found from Eq. (4.14.4). The radiation coefficient from the plate to the glass, $h_{r,p-c}$, is

$$h_{r,p-c} = \frac{\sigma(T_p^2 + T_c^2)(T_p + T_c)}{(1/\epsilon_p) + (1/\epsilon_c) - 1}$$

and the radiation coefficient from the cover to the sky, $h_{r,c-s}$, is

$$h_{r,c-s} = \epsilon_c \sigma(T_c^2 + T_s^2)(T_c + T_s)$$

The cover glass temperature is found by noting that the heat loss from the plate to the cover is the same as from the plate to the surroundings. Therefore

$$T_c = T_p - \frac{U_t(T_p - T_a)}{h_{p-c} + h_{r,p-c}}$$

The procedure is to guess a cover temperature from which h_{p-c}, $h_{r,p-c}$, and $h_{r,c-s}$ are calculated. With these heat transfer coefficients and h_w, the top loss coefficient is calculated. These results are then used to calculate T_c from the above equation. If T_c is close to the initial guess, no further calculations are necessary. Otherwise, the newly calculated T_c is used and the process is repeated.

With an assumed value of the cover temperature of 35 °C, the two radiation coefficients become

$$h_{r,p-c} = 6.44 \ \text{W/m}^2 \ °\text{C}$$

$$h_{r,c-s} = 5.16 \ \text{W/m}^2 \ °\text{C}$$

From Eqs. (4.11.4a) and (4.11.7), the convection coefficient between the plate and glass becomes

$$h_{p-c} = [1 - .0018(50 - 10)] \frac{1.14(30)^{.31}}{(2.54)^{0.070}} = 2.84 \ W/m^2 \ ^\circ C$$

The wind coefficient, from Eq. (4.14.4) is

$$h_w = 5.7 + 3.8(5) = 24.7 \ W/m^2 \ ^\circ C$$

The first estimate of U_t is then

$$U_t = \left(\frac{1}{2.84 + 6.44} + \frac{1}{24.7 + 5.16} \right)^{-1} = 7.08 \ W/m^2 \ ^\circ C$$

The cover glass temperature is

$$T_c = 65 - \frac{7.08(55)}{2.84 + 6.44} = 23 \ ^\circ C$$

With this new estimate of the plate temperature, the various heat transfer coefficients become

$$h_{r,p-c} = 6.10 \ W/m^2 \ ^\circ C$$

$$h_{r,c-s} = 4.84 \ W/m^2 \ ^\circ C$$

$$h_{p-c} = 3.19 \ W/m^2 \ ^\circ C$$

and the second estimate of U_t is

$$U_t = 7.07 \ W/m^2 \ ^\circ C$$

Since the top loss coefficient and the sum of $h_{r,p-c}$ and h_{p-c} did not change significantly from the first calculation, the glass temperature will be essentially 23 °C and the iteration is complete. It is interesting to note that even though the magnitudes of the various heat transfer coefficients that make up the top loss coefficient changed significantly between the two iterations, the magnitude of U_t did not change.

The results of heat loss calculations for four different configurations are given in Figure 7.4.3. The cover temperatures and the heat flux by convection and radiation are shown for one and two glass covers and for selective and nonselective absorber plates. Note that radiation between plates is the dominant mode of heat transfer in the absence of a selective surface. When a selective surface having an emittance of 0.10 is used, convection is the dominant heat transfer mode between the selective surface and the cover, but radiation is still the largest term between the two cover glasses in the two-cover system.

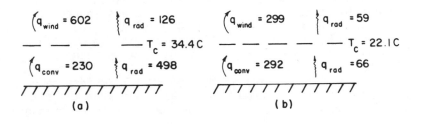

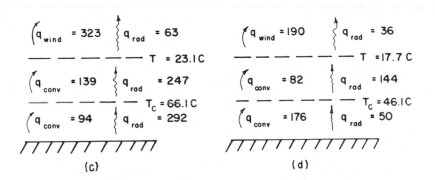

Figure 7.4.3 Temperature distribution and upward heat loss terms for flat plate collectors operating at 100 °C with ambient and sky temperatures of 10 °C, plate spacing of 2.5 cm, tilt of 45° and wind speed of 5 m/s. (All heat flux terms in W/m^2.)
(a) one cover, plate emittance = 0.95; U_t = 8.1 W/m^2 °C.
(b) one cover, plate emittance = 0.10; U_t = 4.0 W/m^2 °C.
(c) two covers, plate emittance = 0.95; U_t = 4.3 W/m^2 °C.
(d) two covers, plate emittance = 0.10; U_t = 2.5 W/m^2 °C.

The heat loss to the environment by convection from the cover glass, due to a 5 m/s wind, is usually about five times the radiation heat loss. Since a 5 m/s wind speed is close to the average for most of the United States, this distribution of heat loss to the surroundings is close to the average conditions. The use of a blackbody radiation sky temperature that is not equal to the air temperature will not significantly affect the overall loss coefficient or the heat loss. For example, the overall loss coefficient for (b) of Figure 7.4.3 is increased from 3.98 to 4.05 W/m^2 °C when the sky temperature is reduced from 10 to 0 °C.

As illustrated by Example 7.4.1, the calculation of the top loss coefficient is a tedious process. To simplify the calculation of collector performance, Figures 7.4.4a through 7.4.4f have been prepared. These figures give the top loss coefficient for one, two, and three glass covers spaced 2.54 cm apart; for ambient temperatures of 40, 10, and -20 °C; for wind speeds of 0, 5, and 10 m/s; for plates having an emittance of 0.95 and 0.10; and for the range of plate temperatures from 10 to 130 °C.

The peculiar behavior of the curves for ambient temperatures below 40 °C is caused by the fact that when the plate temperature is decreased to 40 °C or below, the temperature gradient is reversed and convection between the plates ceases. The mechanism for heat transfer (actually a heat gain) is then by radiation and conduction rather than radiation and convection. Since the conduction resistance is not dependent on the temperature level or difference, except for property variations, the curves exhibit a sudden change in slope.

Even though the top loss coefficients of Figure 7.4.4 are for a plate spacing of 2.54 cm, they can be used for other spacings with little error. The convection heat transfer coefficient between parallel plates is proportional to the plate spacing to the $(3n - 1)$ power where n is the exponent of Eqs. (4.11.3) through (4.11.6), depending upon the tilt angle. This exponent is nearly 1/3, which would eliminate all dependence on plate spacing. In fact, the exponent is 1/3 for the Dropkin and Somerscales correlation, Eq. (4.11.8). Consequently, the dependence of U_t on the plate spacing is very small.

A somewhat more important variable is the tilt (slope), s. Figures 7.4.4a through 7.4.4f were prepared using a tilt of 45°. It is possible to correct for tilt by a method proposed by Klein (1973). In Figure 7.4.5 the ratio of the top loss coefficient at any tilt angle to that at 45° has been plotted as a function of tilt. An equation that represents this graph and can be used to interpolate for other plate emittances is

$$\frac{U_t(s)}{U_t(45)} = 1 - (s - 45)(0.00259 - 0.00144\ \varepsilon_p) \qquad (7.4.8)$$

where s is the tilt in degrees.

The graphs for U_t are convenient for hand calculations but they are difficult to use in computer simulations. The solution to the set of equations, as was done in Example 7.4.1, is time consuming even on a high-speed digital computer since many thousands of solutions may be required. An empirical equation for U_t was as developed by Klein (1973), following the basic procedure of Hottel and Woertz (1942). This new relationship fits the graphs for U_t for plate temperatures between 40 and 130 °C to within ± 0.2 W/m² °C;

@ 45°
$$U_t = \left(\frac{N}{(344/T_p)[(T_p - T_a)/(N + f)]^{0.31}} + \frac{1}{h_w} \right)^{-1}$$

$$+ \frac{\sigma(T_p + T_a)(T_p^2 + T_a^2)}{[\varepsilon_p + 0.0425N(1 - \varepsilon_p)]^{-1} + [(2N + f - 1)/\varepsilon_g] - N}$$

$$(7.4.9)$$

where N = number of glass covers;
$\quad f = (1.0 - 0.04 h_w + 5.0 \times 10^{-4} h_w^2)(1 + 0.058N)$;
$\quad \varepsilon_g$ = emittance of glass (0.88);
$\quad \varepsilon_p$ = emittance of plate;
$\quad T_a$ = ambient temperature (°K);
$\quad T_p$ = plate temperature (°K) (see section 7.8); and
$\quad h_w$ = wind heat transfer coefficient from Eq. (4.14.4).

Equations (7.4.8) and (7.4.9) are convenient both for hand calculations and for calculations on a digital computer. To use either of these empirical relationships or the more complicated exact equations to find U_t, it is necessary to know the mean plate temperature, T_p. A method for estimating T_p is given in section 7.8.

Example 7.4.2 Determine the collector overall loss coefficient (neglecting edge losses) for a single glass cover with the following specifications:

Plate to cover spacing	2.5 cm
Plate emittance	0.95
Ambient temperature	10 °C
Wind speed	5.0 m/s
Back-insulation thickness	5 cm
Insulation conductivity	0.045 W/m °C
Mean plate temperature	65 °C
Collector tilt	53°

$h_w = 5.7 + 3.8 \, v \qquad$ p. 83

$f = (1.0 - 0.04 h_w + 5 \cdot 10^{-4} h_w^2)(1 + 0.058N)$

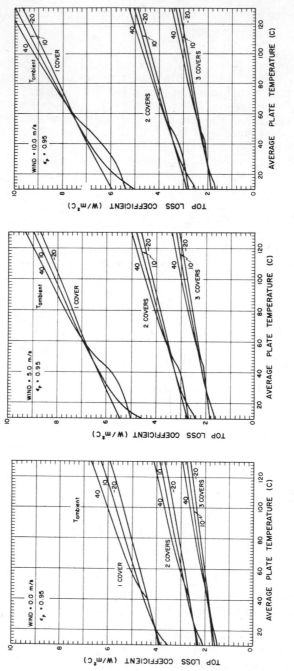

Figure 7.4.4*a,b,c* Top loss coefficient for slope of 45°.

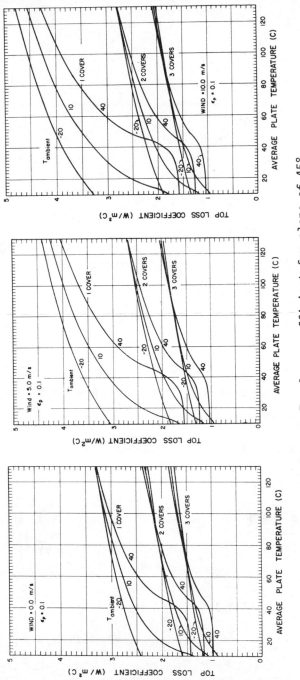

Figure 7.4.4*d,e,f* Top loss coefficient for slope of 45°.

135

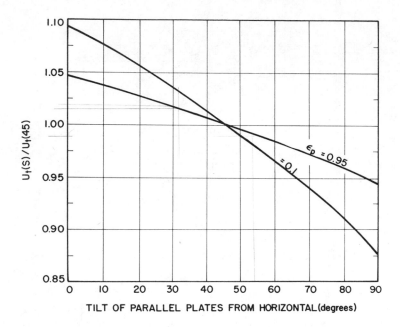

Figure 7.4.5   Angular dependence of top loss coefficient.

From Eq. (4.14.4) the wind coefficient is

$$h_w = 5.7 + 3.8 \times 5.0 = 24.7 \text{ W/m}^2 \text{ °C}$$

From the definition of f below Eq. (7.4.9) we have

$$f = [1.0 - 0.04 \times 24.7 + 5.0 \times 10^{-4}(24.7)^2](1 + 0.058)$$

$$= 0.335$$

From Eq. (7.4.9)*,

$$U_t = \frac{1}{0.310 + 0.040} + \frac{6.84}{1.05 + 1.52 - 1} = 7.2 \text{ W/m}^2 \text{ °C}$$

*The result of the more exact calculation of Example 7.4.1 gives U_t as being equal to 7.1 W/m² °C.

From Eq. (7.4.8) (or Figure 7.4.5) with a collector tilt of 53°,

$$\frac{U_t(53)}{U_t(45)} = 1 - 8(0.0012) = 0.99$$

Therefore, $U_t(53) = 0.99 \times 7.2 = 7.1$ W/m² °C.
The bottom loss coefficient is found from Eq. (7.4.1):

$$U_b = \frac{0.045}{0.05} = 0.9 \text{ W/m}^2 \text{ °C}$$

Finally, the overall loss coefficient, U_L, is found by adding together the top and bottom coefficients:

$$U_L = U_t + U_b = 7.1 + 0.9 = 8.0 \text{ W/m}^2 \text{ °C}$$

Example 7.4.3 For Example 7.4.2, include the effects of edge losses in calculating the overall loss coefficient. Assume the collector bank is 3 m by 10 m or 26 m around.

The collector thickness is 7.5 cm. If the edge insulation is the same as the bottom insulation, the edge loss coefficient times the edge area is

$$U_{edge}A_{edge} = \frac{0.045 \times 0.075 \times 26}{0.05} = 1.76 \text{ W/°C}$$

If we base the loss on the collector frontal area we have

$$U_e = \frac{U_{edge}A_{edge}}{A_{collector}} = \frac{1.76}{30} = 0.59 \text{ W/m}^2 \text{ °C}$$

The edge loss for this 30-m² collector is less than 1% of the top and bottom loss. Note, however, that if this collector were 1 m by 2 m, the edge loss would increase to nearly 3%. Thus, edge losses for large collectors are usually negligible but for small collectors the edge losses may be significant.

The preceding discussion of top loss coefficients was based on glass covers that are opaque to long-wavelength radiation. If a plastic material is used to replace one or more covers, the equation for U_t must be modified to account for some infrared radiation passing directly through the cover. For a single cover that is partially transparent to infrared radiation, the net radiant energy transfer directly between the collector plate and the sky is

$$q_{\text{plate-sky}} = \tau\varepsilon_p \ \sigma(T_p^4 - T_{\text{sky}}^4) \qquad (7.4.10)$$

where τ is the transmittance of the cover for radiation from T_p and from T_{sky} (i.e., we have assumed that the transmittance is independent of source temperature). The top loss coefficient then becomes

$$U_t = \tau\varepsilon_p 4\sigma\bar{T}^3 \ \frac{(T_p - T_s)}{(T_p - T_a)} + \left(\frac{1}{h_{p-c} + h_{r,p-c}} + \frac{1}{h_w + h_{r,c-s}}\right)^{-1}$$

$$(7.4.11)$$

In addition to Eq. (7.4.11), Whillier (1967) has presented top loss coefficients for collector cover systems of one glass cover over one plastic cover, two plastic covers, and one glass cover over two plastic covers.

7.5 TEMPERATURE DISTRIBUTION BETWEEN TUBES AND THE COLLECTOR EFFICIENCY FACTOR

The temperature distribution between two tubes can be derived if we temporarily assume the temperature gradient in the flow direction is negligible. Consider the sheet-tube configuration as shown in Figure 7.5.1. The distance between the tubes is W, the tube diameter is D, and the sheet is thin with a thickness δ. Because the sheet material is a good conductor, the temperature gradient through the sheet is negligible. We will assume the sheet above the bond is at some local base temperature, T_b. The region between the centerline separating the tubes and the tube base can then be considered as a classical fin problem.

The fin, shown in Figure 7.5.2a is of length $(W - D)/2$. An elemental region of width Δx and unit length in the flow direction is shown in Figure 7.5.2b. An energy balance on this element yields

$$S\Delta x + U_L\Delta x(T_a - T) + \left(-k\delta \frac{dT}{dx}\right)\Big|_x - \left(-k\delta \frac{dT}{dx}\right)\Big|_{x + \Delta x} = 0$$

$$(7.5.1)$$

Dividing through by Δx and finding the limit as Δx approaches zero yields

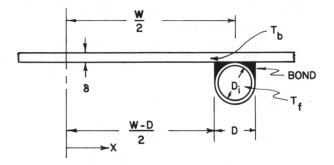

Figure 7.5.1   Sheet and tube dimensions.

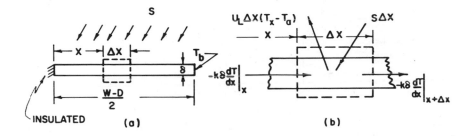

Figure 7.5.2   Energy balance on fin element.

$$\frac{d^2T}{dx^2} = \frac{U_L}{k\delta}\left(T - T_a - \frac{S}{U_L}\right) \tag{7.5.2}$$

The two boundary conditions necessary for this second-order differential equation are symmetry at the centerline and known root
temperature;

$$\frac{dT}{dx}\bigg|_{x = 0} = 0, \quad T\bigg|_{x = (W - D)/2} = T_b \tag{7.5.3}$$

If we define $m^2 = U_L/k\delta$ and $\Psi = T - T_a - S/U_L$, Eq. (7.5.2) becomes

$$\frac{d^2\Psi}{dx^2} - m^2\Psi = 0 \qquad (7.5.4)$$

which has the boundary conditions

$$\frac{d\Psi}{dx}\bigg|_{x=0} = 0, \quad \Psi\bigg|_{x=(W-D)/2} = T_b - T_a - \frac{S}{U_L} \quad (7.5.5)$$

The general solution is then

$$\Psi = C_1 \sinh mx + C_2 \cosh mx \qquad (7.5.6)$$

The constants C_1 and C_2 can be found by substituting the boundary conditions, Eq. (7.5.5), into the general solution. The results are:

$$\frac{T - T_a - S/U_L}{T_b - T_a - S/U_L} = \frac{\cosh mx}{\cosh m(W-D)/2} \qquad (7.5.7)$$

The energy conducted to the region of the tube per unit of length in the flow direction can now be found by evaluating Fourier's law at the fin base:

$$q'_{\text{fin - base}} = -k\delta \frac{dT}{dx}\bigg|_{x=(W-D)/2}$$

$$= \frac{k\delta m}{U_L} [S - U_L(T_b - T_a)] \tanh m \frac{W-D}{2}$$

$$(7.5.8)$$

but $k\delta m/U_L$ is just $1/m$. Equation (7.5.8) accounts for the energy collected on only one side of a tube; therefore, for both sides, the energy collection is

$$q'_{\text{fin - base}} = (W-D)[S - U_L(T_b - T_a)] \frac{\tanh m(W-D)/2}{m(W-D)/2}$$

$$(7.5.9)$$

It is convenient to use the concept of a fin efficiency to rewrite Eq. (7.5.9) as

$$q'_{\text{fin - base}} = (W-D)F[S - U_L(T_b - T_a)] \qquad (7.5.10)$$

where

$$F = \frac{[\tanh m(W - D)/2]}{m(W - D)/2} \tag{7.5.11}$$

The function F is the standard fin efficiency for straight fins with rectangular profile and is plotted in Figure 7.5.3.

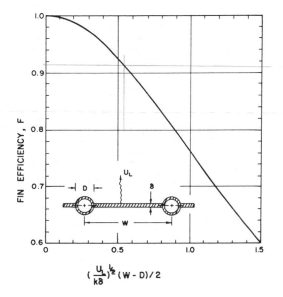

Figure 7.5.3 Fin efficiency for tube and sheet solar collectors.

The useful gain of the collector also includes the energy collected above the tube region. The energy gain for the tube region is

$$q'_{\text{tube}} = D[S - U_L(T_b - T_a)] \tag{7.5.12}$$

and the useful gain for the collector per unit of length in the flow direction becomes

$$q'_u = [(W - D)F + D][S - U_L(T_b - T_a)] \tag{7.5.13}$$

Ultimately, the useful gain from Eq. (7.5.13) must be transferred to the fluid. The resistance to heat flow to the fluid
results from the bond and the fluid to tube resistance. The
useful gain can be expressed in terms of these two resistances
as

$$q'_u = \frac{T_b - T_f}{1/(h_{f,i} \cdot \pi D_i) + 1/C_b} \qquad (7.5.14)$$

where D_i is the inside tube diameter and $h_{f,i}$ is the heat transfer coefficient between the fluid and the tube wall. The bond
conductance, C_b, can be estimated from knowledge of the bond
thermal conductivity, k, the bond average thickness, γ, and the
bond length, b. That is, on a per unit length basis,

$$C_b = \frac{k_b b}{\gamma} \qquad (7.5.15)$$

The bond conductance can be very important in accurately
describing collector performance. Whillier and Saluja (1965)
have shown by experiments that simple wiring or clamping of the
tubes to the sheet results in a significant loss of performance.
They conclude that it is necessary to have good metal-to-metal
contact so that the bond resistance is less than 0.03 m °C/W.
We now wish to eliminate T_b from consideration and obtain
an expression for the useful gain in terms of known dimensions,
physical parameters, and the local fluid temperature. Solving
Eq. (7.5.14) for T_b, substituting it into (7.5.13), and solving
the result for the useful gain, we obtain

$$q'_u = WF'[S - U_L(T_f - T_a)] \qquad (7.5.16)$$

where F', the collector efficiency factor, is

$$F' = \frac{1/U_L}{W\left[\dfrac{1}{U_L[D + (W - D)F]} + \dfrac{1}{C_b} + \dfrac{1}{\pi D_i h_{f,i}}\right]} \qquad (7.5.17)$$

For this and most, but not all geometries, a physical interpretation for the parameter F' becomes clear when it is recognized that the denominator of Eq. (7.5.17) is the heat transfer
resistance from the fluid to the ambient air. This resistance
will be given the symbol $1/U_0$. The numerator is the heat transfer resistance from the absorber plate to the ambient air. F'
is thus the ratio of these two heat transfer coefficients

$$F' = \frac{U_0}{U_L} \qquad\qquad (7.5.18)$$

Another interpretation for F' results from Eq. (7.5.16). At a particular location, F' represents the ratio of the useful energy gain to the useful energy gain if the collector absorbing surface had been at the local fluid temperature.

The collector efficiency factor is essentially a constant for any collector design and fluid flow rate. The ratio of U_L to C_b, the ratio of U_L to $h_{f,i}$, and the fin efficiency parameter F are the only variables appearing in Eq. (7.5.17) that may be functions of temperature. For most collector designs F is the most important of these variables in determining F', but it is not a strong function of temperature.

The evaluation of F' from Eq. (7.5.17) is not a difficult task. However, to illustrate the effects of various design parameters on the magnitude of F', Figure 7.5.4 has been prepared. Three values of the overall heat transfer coefficient U_L were chosen (2, 4, and 8 W/m^2 °C) which correspond approximately to three, two and one cover nonselective collectors with a wind of 5 m/s. See Figure 7.4.4 for other combinations that will yield these same overall loss coefficients. Instead of selecting various plate materials, the curves were prepared for various values of $(k\delta)$, the product of the plate thermal conductivity, and plate thickness. For a copper plate, 1 mm thick, $k\delta$ is equal to 0.4 W/°C; for a steel plate 0.1 mm thick, $k\delta$ is equal to 0.005 W/°C. Thus, the probable range of $k\delta$ is from 0.005 to 0.4.

The heat transfer coefficient inside the tubes was assumed to be either 300 W/m^2 °C, or 1500 W/m^2 °C. As expected, the collector efficiency factor decreases with increased tube center to center distances and increases with increases in both material thicknesses and thermal conductivity. Increasing the overall loss coefficient decreases F' while increasing the fluid to tube hat transfer coefficient increases F'.

7.6 TEMPERATURE DISTRIBUTION IN FLOW DIRECTION

The useful gain per unit of flow length as calculated from Eq. (7.5.16) is ultimately transferred to the fluid. The fluid enters the collector at temperature $T_{f,i}$ and increases in temperature until at exit it is $T_{f,o}$. Referring to Figure 7.6.1, we can express an energy balance on the fluid flowing through a section

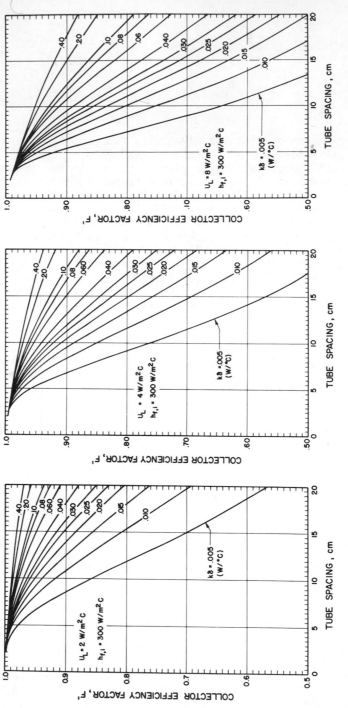

Figure 7.5.4*a,b,c* Collector efficiency factor F' versus tube spacing for 2 cm diameter tubes for various conditions.

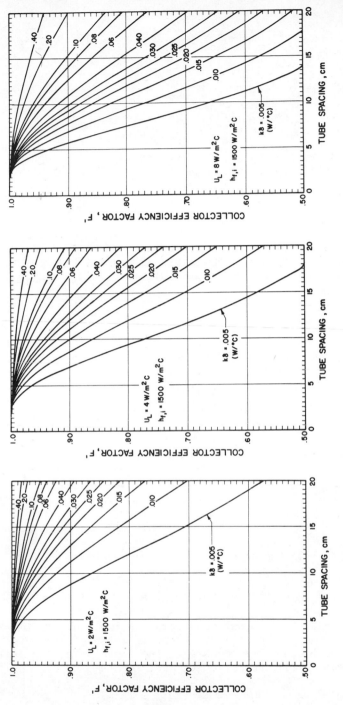

Figure 7.5.4d,e,f Collector efficiency factor F' versus tube spacing for 2 cm diameter tubes for various conditions.

145

of one pipe of length Δy as

$$\dot{m}C_p T_f\Big|_y - \dot{m}C_p T_f\Big|_{y + \Delta y} + \Delta y q'_u = 0 \qquad (7.6.1)$$

Dividing through by Δy, finding the limit as Δy approaches zero, and substituting Eq. (7.5.16) for q'_u, we thus obtain

$$\dot{m}C_p \frac{dT_f}{dy} - WF'[S - U_L(T_f - T_a)] = 0 \qquad (7.6.2)$$

If we assume that F' and U_L are independent of position, then the solution for the temperature at any position y (subject to the condition that the inlet fluid temperature is $T_{f,i}$) is

$$\frac{T_f - T_a - S/U_L}{T_{f,i} - T_a - S/U_L} = e^{-[U_L WF'y/\dot{m}C_p]} \qquad (7.6.3)$$

If the collector has a length L in the flow direction, then the outlet fluid temperature, $T_{f,o}$ is found by substituting L for y in Eq. (7.6.3).

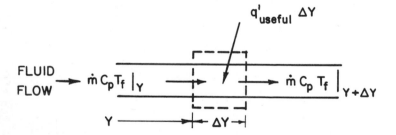

Figure 7.6.1 Energy balance on fluid element.

7.7 COLLECTOR HEAT REMOVAL FACTOR AND FLOW FACTOR

It is convenient to define a quantity that relates the actual useful energy gain of a collector to the useful gain if the whole collector surface were at the fluid inlet temperature. Mathematically, the collector heat removal factor, F_R, is then

$$F_R = \frac{GC_p(T_{f,o} - T_{f,i})}{[S - U_L(T_{f,i} - T_a)]} \qquad (7.7.1)$$

where G is the flow rate per unit of collector area.

The collector heat removal factor can be expressed as

$$F_R = \frac{GC_p}{U_L}\left[\frac{T_{f,o} - T_{f,i}}{S/U_L - (T_{f,i} - T_a)}\right]$$

$$= \frac{GC_p}{U_L}\left[\frac{(T_{f,o} - T_a - S/U_L) - (T_{f,i} - T_a - S/U_L)}{S/U_L - (T_{f,i} - T_a)}\right]$$

$$(7.7.2)$$

or

$$F_R = \frac{GC_p}{U_L}\left[1 - \frac{S/U_L - (T_{f,o} - T_a)}{S/U_L - (T_{f,i} - T_a)}\right] \qquad (7.7.3)$$

which, from Eq. (7.6.3) with $y = L$, can be expressed as

$$F_R = \frac{GC_p}{U_L}\left(1 - e^{-[U_L F'/GC_p]}\right) \qquad (7.7.4)$$

To present Eq. (7.7.4) graphically, it is convenient to define a new variable, F'', which is equal to F_R/F'. This collector flow factor is a function of the single variable $U_L F'/GC_p$ (or U_o/GC_p) and is shown in Figure 7.7.1. For a full appreciation of the significance of this graph, it is convenient to re-examine Eq. (7.7.1) but in a slightly different form:

$$Q_u = A_c F_R[S - U_L(T_{f,i} - T_a)] \qquad (7.7.5)$$

where Q_u is the total useful energy gain of the collector. With this equation, the useful energy gain is calculated as a function of the inlet fluid temperature. This is a convenient representation when analyzing solar energy systems since the inlet fluid temperature is usually known. However, it must be remembered that losses based on the inlet temperature are too small since losses occur all along the collector and the fluid has an ever increasing temperature in the flow direction. The effect of the multiplier, F_R, is to reduce the calculated useful energy gain from what it would have been had the whole collector been at $T_{f,i}$ to what it actually is using a fluid that increases in temperature as it flows through the collector. As the mass flow rate through the collector increases, the temperature rise through the collector decreases. This causes lower

losses and a corresponding increase in the useful energy gain
since the average collector temperature is lower. This increase
in useful energy gain is reflected by an increase in the collector heat removal factor F_R as the mass flow rate increases.
Note that the heat removal factor F_R can never exceed the collector efficiency factor F'. As the flow rate becomes very
large, the temperature rise from inlet to outlet decreases
toward zero but the temperature of the absorbing surface will
still be higher than the fluid temperature. This temperature
difference is accounted for by the collector efficiency factor,
F'.

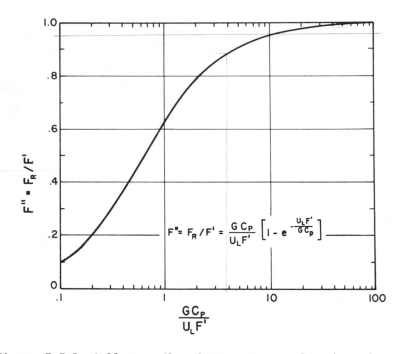

Figure 7.7.1 Collector flow factor F'' as a function of
GC_p/U_LF'.

Example 7.7.1 Calculate the daily efficiency of a solar
collector operating in Madison on January 10. Assume the
collector is tilted toward the equator at an angle of 53°
(i.e., 10° greater than the latitude). The hourly radiation and ambient air temperature are as given below.

For the collector, assume U_L is 8.0 W/m² °C, aluminum fin
and tube type construction, tube center to center distance
is 15 cm, fin thickness is 0.05 cm, tube diameter is 1.6
cm, fluid to tube heat transfer coefficient is 1500 W/m²
°C with negligible bond resistance, cover transmittance
for solar radiation is 0.88 and is independent of direction, solar absorptance of absorbing plate is 0.95 and independent of direction, collector width is 1 m and length
is 2 m, water flow rate is 0.02 kg/s, and water inlet temperature is constant and equal to 60 °C.

TABLE 7.7.1 Data for January 15 at a Latitude of 43°.

Time, Hr	H, W/m²	T, °C
7-8	14	-1
8-9	76	0
9-10	174	2
10-11	306	4
11-12	449	10
12-13	461	10
13-14	389	8
14-15	275	8
15-16	127	6
16-17	21	4

The efficiency of the collector for each hour of operation can be found from

$$\eta_{hour} = \frac{Q_u}{HRA_c}$$

where HR is the radiation on the collector surface, Q_u is
the useful energy gain for the hour, and A_c is the collector area. The daily efficiency is not the average hourly
efficiency but must be calculated from

$$\eta_{day} = \frac{\Sigma Q_u}{A_c \Sigma HR}$$

where ΣQ_u is the total useful gain for the day and ΣHR is the total incident solar radiation. In order to calculate Q_u, it is necessary to determine the factors F and F' and F_R from Eq . (7.5.11), (7.5.17), and (7.7.4), respectively. From the definition of m in section 7.5,

$$m = \left(\frac{U_L}{k\delta}\right)^{\frac{1}{2}} = \left[\frac{8.0}{210 \times 5 \times 10^{-4}}\right]^{\frac{1}{2}} = 8.73$$

The half-length between the tubes is

$$\frac{W - D}{2} = \frac{0.15 - 0.016}{2} = 0.067$$

From Eq. (7.5.11),

$$F = \frac{\tanh\ (8.73 \times 0.067)}{(8.73 \times 0.067)} = 0.900$$

The factor F' is found from Eq. (7.5.17):

$$F' = \frac{1/8}{0.15[1/\{8(0.01 + 0.134 \times 0.90\} + 1/(\pi \times 0.01 \times 1500)]}$$

$$= 0.897$$

and finally, F_R is found from Eq. (7.7.4) with $G = 0.02/2 \times 1 = 0.01$ Kg/s m^2.

$$F_R = \frac{0.01 \times 4.187 \times 10^3}{8}\left[1 - \exp\left(- \frac{8 \times 0.897}{0.01 \times 4.187 \times 10^3}\right)\right] = 0.824$$

The transmittance-absorptance product of this cover and absorber plate system is found using Eq. (6.3.1):

$$(\tau\alpha) = \frac{(0.88)0.95}{1 - (1 - 0.95)0.16} = 0.84$$

To evaluate $HR(\tau\alpha)$ for each hour, it is necessary to calculate the factor R, which converts horizontal radiation to radiation on the tilted collector for each hour.* From Eq. (3.6.2) with $\delta = -22°$, $\phi = 43°$, and $s = 53°$

*In this example we have used the correction factor R, calculated for beam radiation only.

$$\cos \theta_T = \cos (43 - 53) \cos (-22) \cos \omega + \sin (43 - 53) \sin (-22)$$

$$= 0.914 \cos \omega + 0.065$$

$$\cos \theta_z = \cos (43) \cos (-22) \cos \omega + \sin (43) \sin (-22)$$

$$= 0.678 \cos \omega - 0.256$$

The following work table (Table 7.7.2) gives R and other quantities for the midpoint of the hours indicated.

TABLE 7.7.2   Work table for Example 7.7.2.

Time	T_{amb}, °C	H, W/m^2	ω	$\cos \theta_T$	$\cos \theta_z$	R	HR, W/m^2	$HR\tau\alpha$, W/m^2	$U_L(T - T_{amb})$ W/m^2	q_u W/m^2
7-8	1	14	+67.5	0.415	0.006	--*	--	--	--	--
8-9	0	76	+52.5	0.620	0.156	3.98	303	254	480	0
9-10	2	174	+37.5	0.789	0.281	2.80	486	409	464	0
10-11	4	306	+22.5	0.910	0.369	2.44	746	627	448	147
11-12	10	449	+7.5	0.971	0.417	2.33	1050	879	400	393
12-13	10	461	-7.5	0.971	0.417	2.33	1075	902	400	412
13-14	8	389	-22.5	0.910	0.369	2.44	948	797	416	312
14-15	8	275	-37.5	0.789	0.281	2.80	770	647	416	189
15-16	6	127	-52.5	0.620	0.156	3.98	505	424	432	0
16-17	4	21	-67.5	0.415	0.006	--*	--	--	--	--
Totals							5883			1453

$$\eta = \frac{\Sigma q_u}{\Sigma HR} = \frac{1453}{5883} = 24.7\%$$

*The sun rises at 7:30 and sets at 4:30 at this latitude. The ratio R at sunrise and sunset approaches infinity under these conditions. In reality, the incident radiation is small and will be neglected.

7.8 MEAN PLATE TEMPERATURE

To evaluate collector performance, it is necessary to know the overall loss coefficient U_L. However, U_L is a function of plate

temperature and an iterative approach becomes necessary. The
mean fluid temperature can be found by integrating Eq. (7.6.3)
from zero to L:

$$T_{f,m} = \frac{1}{L} \int_0^L T_{f,y} \, dy \tag{7.8.1}$$

Performing this integration and substituting F_R from Eq. (7.7.4)
and Q_u from Eq. (7.7.5), the mean fluid temperature was shown
by Klein (1973) to be

$$T_{f,m} = T_{f,i} + \frac{Q_u/A}{U_L F_R} \left[1 - \frac{F_R}{F'} \right] \tag{7.8.2}$$

The mean plate temperature will always be greater than the
mean fluid temperature due to the heat transfer resistance between the absorbing surface and the fluid. This temperature
difference is usually small for liquid systems but may be significant for air systems.

The temperature difference between the absorber plate and
the fluid will not be constant along the flow direction due to
changes in the collector heat loss. However, as an approximation, the mean fluid temperature and the mean plate temperature
are related by

$$T_{p,m} - T_{f,m} = Q_u R_{p-f} \tag{7.8.3}$$

where R_{p-f} is the heat transfer resistance between the plate
and the fluid. For both liquid and air systems the major heat
transfer resistance between the plate and the fluid is the convection heat transfer coefficient. For a fin and tube collector this assumes both the bond conductance and the fin efficiency
are high. Thus, R_{p-f} for liquid flowing in tubes is given by
$1/(h_{f,i} \pi D_i n L)$ where n and L are the number of tubes and their
length, respectively. For a simple air system (as shown in
Figure 7.12.1e) this resistance is just the reciprocal of the
plate to air heat transfer coefficient times the collector area.

Equations (7.8.2) and (7.8.3) can be solved in an iterative
manner with Eq. (7.4.9). First an estimate of the mean plate
temperature is made from which U_L is calculated. With approximate values of F_R, F', and Q_u, a new mean plate temperature is
obtained from Eqs. (7.8.2) and (7.8.3) and used to find a new
value of the top loss coefficient. The new value of U_L is used

to refine F_R, and F', and the process is repeated. With a reasonable initial guess, this iterative process is seldom necessary.

  Example 7.8.1 Find the mean plate temperature for Example 7.7.1.

  With U_L equal to 8.0 W/m^2 °C, $F' = 0.897$, $F_R = 0.824$, and $q_u = 300$ W/m^2, we have from Eq. (7.8.2)

$$T_{f,m} = 60 + \frac{300}{8 \times 0.824}\left[1 - \frac{0.824}{0.897}\right] = 64 \text{ °C}$$

The value for Q_u/A (300 W/m^2) used in the calculation for $T_{f,m}$ is approximately the average useful energy gain.

The mean plate temperature is found from Eq. (7.8.3):

$$T_{p,m} = 64 + \frac{300 \times 2 \times 1}{1500 \times \pi \times 0.016 \times 2 \times 1/0.15} = 65 \text{ °C}$$

This is the value of mean plate temperature that was initially used in Example 7.4.2 to find U_L. Note that the mean plate temperature changes slightly as q_u changes throughout the day, but the influence on U_L will be small.

7.9   EFFECTIVE TRANSMITTANCE-ABSORPTANCE PRODUCT

In section 6.3, the product of cover transmittance times plate solar absorptance was discussed. In section 7.4 the expressions for U_L were derived assuming that the cover system did not absorb solar radiation. In order to maintain the simplicity of Eq. (7.7.5) and account for the reduced losses due to absorption of solar radiation by the glass, an effective transmittance-absorptance product will be introduced.
  All of the solar radiation that is absorbed by a cover system is not lost, since this absorbed energy tends to increase the cover temperature and consequently reduce the losses from the plate. Let us consider the thermal network of a single-cover system as shown in Figure 7.9.1. The solar energy absorbed by the cover is $HR(1 - \tau_a)$, where τ_a is the transmittance considering only absorption from Eq. (6.2.2). The loss for (a), without absorption, is $U_1(T_p - T_c)$ and the loss for (b), with absorption, is $U_1(T_p - T_c')$. Here we have assumed that the small amount of absorption in the cover and consequent increased cover temperature does not change the magnitudes of U_1 and U_2. The

difference, D, in the two loss terms is

$$D = U_1[(T_p - T_c) - (T_p - T_c')] \qquad (7.9.1)$$

The temperature difference $(T_p - T_c)$ can be expressed as

$$(T_p - T_c) = \frac{(T_p - T_a)U_L}{U_1} \qquad (7.9.2)$$

where U_L is the overall loss coefficient and is equal to $U_1U_2/(U_1 + U_2)$. For simplicity, we have assumed the overall loss coefficient and the top loss coefficient are identical.

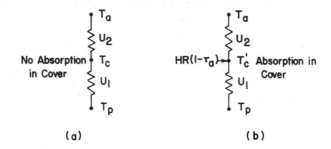

(a)   (b)

Figure 7.9.1 Thermal network for top losses for a single cover collector with and without absorption in the cover.

The temperature difference $(T_p - T_c')$ can be expressed as

$$(T_p - T_c') = \frac{U_2(T_p - T_a) - HR(1 - \tau_a)}{U_1 + U_2} \qquad (7.9.3)$$

Therefore

$$D = (T_p - T_a)U_L - \frac{U_1U_2(T_p - T_a)}{U_1 + U_2} + \frac{HR(1 - \tau_a)U_1}{U_1 + U_2} \qquad (7.9.4)$$

or

$$D = HR(1 - \tau_a)\frac{U_L}{U_2} \qquad (7.9.5)$$

The quantity D represents the reduction in collector losses due to absorption in the cover but can be considered an additional input in the collector equation. The useful gain of a collector is then

$$q_u = F_R \left\{ HR \left[(\tau\alpha) + (1 - \tau_a) \frac{U_L}{U_2} \right] - U_L (T_{f,i} - T_a) \right\} \quad (7.9.6)$$

By defining the quantity $(\tau\alpha) + (1 - \tau_a)U_L/U_2$ as the effective transmittance-absorptance product, the simplicity of Eq. (7.7.5) can be maintained. For this one-cover system

$$(\tau\alpha)_e = (\tau\alpha) + (1 - \tau_a) \frac{U_L}{U_2} \quad (7.9.7)$$

A general analysis for a cover system of n identical plates yields

$$(\tau\alpha)_e = (\tau\alpha) + (1 - \tau_a) \sum_{i=1}^{n} a_i \tau^{i-1} \quad (7.9.8)$$

where a_i is the ratio of the overall loss coefficient to the loss coefficient from the ith cover to the surroundings and τ_a is the transmittance of a single cover from Eq. (6.2.2). This equation was derived assuming that the transmittance to the ith cover could be approximated by the transmittance of a single cover raised to the $i - 1$ power.

For a cover system composed of different materials (e.g., a combination of glass and plastic) the effective transmittance-absorptance product is

$$(\tau\alpha)_e = (\tau\alpha) + (1 - \tau_{a,1})a_1 + (1 - \tau_{a,2})a_2\tau_1 + (1 - \tau_{a,3})a_3\tau_2$$
$$+ \dots \quad (7.9.9)$$

where τ_i is the transmittance of the cover system above the $i + 1$ cover and $\tau_{a,i}$ is the transmittance due to absorption of the ith cover.

The values of a_i actually depend upon the plate temperature, ambient temperature, plate emittance, and wind speed. Table 7.9.1 gives values of a_i for one, two, and three covers and for plate emittances of 0.95, 0.50, and 0.10. These values

were calculated using a wind speed of 5 m/s, a plate temperature of 100 °C, and an ambient air and sky temperature of 10 °C.
The dependence of a_i on temperature is minor and can be neglected. The dependence of a_i on wind speed may be significant.
For example, a_1 and a_2 are 0.09 and 0.60 for a two-cover collector having a plate emittance of 0.95 with a wind speed of 10 m/s.
However, the sum of a_1 and a_2 for the 5 m/s wind is 0.77 and
for the 10 m/s wind is 0.69. In other words, using the 5 m/s
values of a_i, a 10% error is made in evaluating the amount of
the radiation that is absorbed in the glass that contributes
the useful gain. Since the total amount absorbed by the glass
must be small, a 10% error is not significant. The angular dependence of $(\tau\alpha)_e$ can be evaluated using the proper angular dependence of $(\tau\alpha)$, τ_a, and τ.

TABLE 7.9.1 Constants for Use in Eqs. (7.9.8) and (7.9.9).

Covers	a_i	$\varepsilon_p = 0.95$	$\varepsilon_p = 0.50$	$\varepsilon_p = 0.10$
1	a_1	0.27	0.21	0.13
2	a_1	0.15	0.12	0.09
	a_2	0.62	0.53	0.40
3	a_1	0.14	0.08	0.06
	a_2	0.45	0.40	0.31
	a_3	0.75	0.67	0.53

7.10   EFFECTS OF DUST AND SHADING

The effects of dust and shading are difficult to generalize.
The data of Dietz (1963) shows that at the angles of interest
(0 to 50°) the influence of dirt can be as high as 5%. From
long-term experiments on collectors in the Boston area, Hottel
and Woertz (1942) found that collector performance decreased
about 1% due to dirty glass. In a recent rainless 30 day experiment in India, Garg (1974) found that dust reduced the
transmittance by an average of 8% for glass tilted at 45°.
For design purposes without extensive tests, it is suggested
that radiation absorbed by the plate be reduced by a factor of
$(1 - d)$ where d is 0.02 to account for dust.
    Shading effects can also be significant. Whenever the
angle of incidence is not normal, some of the structure will

intercept solar radiation. Some of this radiation will be reflected to the absorbing plate if the sidewalls are of a high-
reflectance material. Hottel and Woertz recommend that the
radiation absorbed by the plate be reduced by 3% to account for
shading effects, if the net glass area is used in all calculations. The net glass area accounts for the blockage by the supports for the glass. These supports may be a significant contribution to shading losses in some designs, and it may be necessary to do a more elaborate analysis.

Example 7.10.1 In example 7.7.1, the effects of dust,
shading, angle of incidence, and absorption by the cover
were all neglected. Assume that the cover is constructed
from glass $(KL = 0.0370)$ as given in the middle curves of
Figure 6.2.1 and reevaluate the performance. Assume α
does not depend on angle, and n is 1.526.

Since the dust effect is essentially 2% and the shading
effect is nearly 3%, the problem reduces to evaluating the
effective transmittance-absorptance product at the various
times. The transmittance is found from Figure 6.2.1, $(\tau\alpha)$
from Eq. (6.3.1), and $(\tau\alpha)_e$ from Eq. (7.9.8).

Column 1 of the following work table (Table 7.10.1) is
found from Example 7.7.1. Column 2 is found from Figure
6.2.1. Column 3 is found using Eq. (6.3.1) with ρ_d of
0.16.

$$(\tau\alpha) = \frac{\tau(\theta)\alpha}{1 - (1 - \alpha)\rho_d} = 0.96\ \tau(\theta)$$

Column 4 is obtained from Eq. (7.9.8):

$$(\tau\alpha)_e = (\tau\alpha) + 0.27(1 - \tau_a)$$

where τ_a is calculated from Equation 6.2.2.

Column 5 is obtained from $HR(\tau\alpha)_e(1 - d)(1 - s)$ where HR
is from Example 7.7.1, $(1 - d)$ is 0.98, and $(1 - s)$ is
0.97. Therefore

$$s = HR(\tau\alpha)_e(1 - d)(1 - s) = 0.951\ HR(\tau\alpha)_e$$

Column 6 is obtained by subtracting $U_L(T - T_{amb})$ from
column 5. Finally, column 7 is obtained by multiplying
6 by F_R, and 8 by dividing 7 by HR.

TABLE 7.10.1

Time	θ_T	τ	$(\tau\alpha)$	$(\tau\alpha)_e$	S W/m^2	$q_u/F_r,$ W/m^2	$q_u,$ W/m^2	η_{hour}
	1	2	3	4	5	6	7	8
7-8	65	0.73	0.69	0.71	--	--	--	--
8-9	51	0.84	0.80	0.82	236	0	0	0
9-10	38	0.86	0.82	0.83	384	0	0	0
10-11	24	0.88	0.84	0.85	603	155	128	0.17
11-12	14	0.88	0.84	0.85	849	449	370	0.35
12-13	14	0.88	0.84	0.85	869	469	386	0.36
13-14	24	0.88	0.84	0.85	766	350	288	0.30
14-15	38	0.86	0.82	0.83	608	192	158	0.21
15-16	51	0.84	0.80	0.82	394	0	0	0
16-17	65	0.73	0.69	0.71	--	--	--	--
Total							1330	

$$\eta_{day} = \frac{1330}{5883} = 22.6\%$$

Note for that this particular example, all incident angles were near normal whenever there was significant useful gain. This is a common occurrence in flat-plate collectors.

7.11 HEAT CAPACITY EFFECTS IN FLAT-PLATE COLLECTORS

The operation of most solar energy systems is inherently transient; there is no such thing as steady-state operation when one considers the transient nature of the driving forces. This observation has led to a numerical study by Klein, Duffie, and Beckman (1973) into the effects of collector heat capacity on collector performance. The effects can be regarded in two distinct parts; one part is due to the heating of the collector

from its early morning low-temperature to its final operating
temperature in the afternoon. The second part is due to intermittent behavior during the day whenever the driving forces
such as solar radiation and wind change rapidly.

Klein et al. showed that the daily morning heating of the
collector results in a loss that can be significant but is negligible for many situations. For example, the radiation on the
collector of Example 7.10.1 before 10:00 was about 620 W/m^2.
The calculated losses exceeded this value during this time
period because these calculated losses assumed that the fluid
entering the collector was at 60 °C. In reality, no fluid
would be circulating under these conditions and the absorbed
solar energy would heat the collector without reducing the useful energy gain.

The amount of preheating that will occur in a given collector can be calculated by solving the transient energy balance
equations for the various parts of the collector. Even though
these equations can be developed to almost any desired degree
of accuracy, the driving forces such as solar radiation, wind
speed, and ambient temperature are usually known only at hour
intervals. This means that any predicted transient behavior
between the hour intervals can only be approximate, even with
extensive analysis. Consequently, a simplified analysis is
warranted to determine if more detailed analysis is desirable.

To illustrate the method, consider a single-cover collector. We will assume the absorber plate, water in the tubes, and
back-insulation are all at the same temperature. We also assume
that the cover is at a single temperature, but different than
the plate. An energy balance on the collector absorber plate,
water, and back-insulation yields

$$(mC)_p \frac{dT_p}{d\tau} = A_c [S + U_1(T_c - T_p)] \qquad (7.11.1)$$

when the subscripts c and p represent cover and plate and U_1 is
the loss coefficient from the plate to the cover. An energy
balance on the cover yields

$$(mC)_c \frac{dT_c}{d\tau} = A_c[U_1(T_p - T_c) + U_2(T_a - T_c)] \qquad (7.11.2)$$

where U_2 is the loss coefficient from the cover to the ambient
air and T_a is the ambient temperature. It is possible to solve
these two equations simultaneously; however, a great simplification occurs if we assume $(T_c - T_a)/(T_p - T_a)$ remains constant
at its steady-state value. In other words, if we assume that

the following relationship holds:

$$U_2(T_c - T_a) = U_L(T_p - T_a) \qquad (7.11.3)$$

where U_L is the overall loss coefficient, $U_1 U_2/(U_1 + U_2)$, and if we differentiate Eq. (7.11.3), assuming T_a is a constant, we have

$$\frac{dT_c}{d\tau} = \frac{U_L}{U_2} \frac{dT_p}{d\tau} \qquad (7.11.4)$$

If we now add Eq. (7.11.1) to (7.11.2) and use (7.11.4), we obtain the following differential equation for the plate temperature:

$$\left[(mC)_p + \frac{U_L}{U_2} (mC)_c \right] \frac{dT_p}{d\tau} = A_c[S - U_L(T_p - T_a)] \qquad (7.11.5)$$

The term in the square brackets represents an effective heat capacity of the collector. By the same reasoning, the effective heat capacity of a collector with n covers would be

$$(mC)_e = (mC)_p + \sum_{i=1}^{n} a_i (mC)_{c,i} \qquad (7.11.6)$$

where a_i is the ratio of overall loss coefficient to the loss coefficient from the cover in question to the surroundings. This is the same quantity as presented in Table 7.9.1.

The simplification introduced through the use of Eq. (7.11.3) is very significant in that the problem of determining heat capacity effects has been reduced to solving one differential equation instead of $n + 1$ differential equations. The error introduced by this simplification is difficult to assess for all conditions without solving the set of differential equations, but for the example to follow the error is very small. (A worst case analysis would be to assume all the a_i's are equal to unity, but this greatly overestimates the heat capacity loss.)

If we assume that S and T_a remain constant for some period, say one hour, the solution to Eq. (7.11.5) is

$$\frac{S - U_L(T_p - T_a)}{S - U_L(T_{p,\text{initial}} - T_a)} = \exp\left(-\frac{A_c U_L \tau}{(mC)_e} \right) \qquad (7.11.7)$$

The collector plate temperature T_p can be evaluated at the end of each time period by knowing S, U_L, T_a and the collector plate temperature at the beginning of the time period. Repeated application of Eq. (7.11.7) for each hour before the collector actually operates serves to estimate the collector temperature as a function of time. An estimate of the reduction in useful gain can then be obtained by multiplying the collector effective heat capacity by the temperature rise necessary to bring the collector to its initial operating temperature.

A similar loss occurs due to collector heat capacity whenever the final average collector temperature in the afternoon exceeds the initial average temperature. This loss can be easily estimated by multiplying collector effective heat capacity times this temperature difference.

Finally, Klein et al. showed that the effects of intermittent sunshine, wind speed and ambient air temperature were always negligible for normal collector construction.

Example 7.11.1 For the collector described in Example 7.10.1, estimate the reduction in useful energy gain due to heat capacity effects.

Since the collector operates with a constant inlet temperature, only the early morning heating will influence the useful gain.

The collector heat capacity includes the glass, plate, water in the tubes, and insulation. If the glass is 0.32 cm thick, the capacity of the glass is

$$1 \text{ m} \times 2 \text{ m} \times 0.0032 \text{ m} \times 2500 \frac{\text{kg}}{\text{m}^3} \times \frac{1 \text{ kJ}}{\text{kg} - {}^\circ\text{C}} = 16 \text{ kJ}/{}^\circ\text{C}$$

For the plate, water and insulation, the heat capacities are 3, 4, and 3 kJ/°C, respectively. The effective collector capacity is then 10 + 0.27 (16) or 14.3 kJ/°C. From Eq. (7.11.7), the collector temperature at the end of the period from 8:00 to 9:00 a. m. with an initial collector temperature equal to the ambient temperature is

$$T_p = T_a + \frac{S}{U_L}\left[1 - \exp\left(-\frac{A_c U_L \tau}{(mC)_e}\right)\right]$$

$$= 0 + \frac{236}{8}\left[1 - \exp\left(-\frac{2 \times 8 \times 3600}{14300}\right)\right]$$

$$= 29 \ {}^\circ\text{C}$$

For the second hour period, the initial temperature is
39 °C and the temperature at 10:00 a. m. becomes

$$
T_p = T_a + \frac{S}{U_L} - \left[\frac{S}{U_L} - (T_i - T_a) \right] \exp\left(- \frac{A_c U_L}{(mC)_e} \right)
$$

$$
= - 2 + \frac{384}{8} - \left[\frac{384}{8} - 29 - 2 \right] \exp\left(- \frac{2 \times 8 \times 3600}{14300} \right)
$$

$$
= 50 \text{ °C}
$$

By 10:00 a. m., the collector has been heated to within 10 °C of its operating temperature of 60 °C. The reduction in useful gain is the energy to heat the effective
mass times heat capacity of the collector the last 10 °C,
or 143 kJ. Note that this collector design responds quick-
ly to the various changes as the exponential term in the
preceding calculations was negligible. (The collector
"time constant" is $(mC)_e/A_c U_L$, which is about 15 min.)

7.12   OTHER COLLECTOR GEOMETRIES

In the preceding sections, we have considered only one basic
collector design: a sheet and tube solar water heater. There
are many different designs of flat-plate collectors, but fortunately it is not necessary to develop a completely new analysis for each situation. Hottel and Whillier (1958), Whillier
(1967), and Bliss (1959) have shown that the generalized relationships developed for the tube and sheet case apply to most
collector designs. It is necessary to derive the appropriate
form of the collector efficiency factor, F', and Eqs. (7.7.4)
and (7.7.5) can be used to predict the thermal performance.
Under some circumstances, the loss coefficient U_L will have to
be modified slightly.
    Figure 7.12.1 shows ten different liquid and gas collector
designs that have been studied. Also on this figure are equations for the collector efficiency factors that have been derived for these geometries. For (h) and (i), the Löf overlapped
glass plates and the matrix air heater, the analyses to date have
not put the results in a generalized form. For these two situations, it is necessary to resort to numerical techniques for
analysis. Selcuk (1971) has analyzed the overlapped glass plate
system, and Hamid and Beckman (1971), and Chiou et al. (1965)
have studied the matrix-type air heaters.

A somewhat unconventional design is (j), the Speyer (1965) collector, which uses an evacuated glass tube for the cover. Because of the circular geometry, it is possible to evacuate the system and consequently reduce the convection heat loss. With a selective absorbing surface and low emittance surfaces on the back side, the radiation loss can be reduced to relatively small values.

In order to illustrate the procedure for deriving F' and U_L for an air heater, we will derive the equations for (d) of Figure 7.12.1. A schematic of the collector and thermal network are shown in Figure 7.12.2. For this derivation, we will assume back losses are negligible.

At some location along the flow direction the absorbed solar energy heats up the plate to a temperature T_p. Energy is transferred from the plate to the fluid at T_f through the convection heat transfer coefficient h_2 and to the bottom of the cover glass by radiation. Energy is transferred to the cover glass from the fluid through heat transfer coefficient h_1 and, finally, energy is lost to the ambient air through the combined convection and radiation coefficient U_t. Note that U_t can account for multiple covers.

Energy balances on the cover, the plate, and the fluid yield the following equations:

$$U_t(T_a - T_c) + h_r(T_p - T_c) + h_1(T_f - T_c) = 0 \quad (7.12.1)$$

$$S + h_2(T_f - T_p) + h_r(T_c - T_p) = 0 \quad (7.12.2)$$

$$h_1(T_c - T_f) + h_2(T_p - T_f) = q_u \quad (7.12.3)$$

The above three equations need to be solved so that the useful gain is expressed as a function of U_t, h_1, h_2, h_r, T_f, and T_a. In other words, T_p and T_c must be eliminated. The algebra is somewhat tedious and only a few intermediate steps will be given. Solving the first two equations for $(T_p - T_f)$ and $(T_c - T_f)$, we obtain

$$(T_p - T_f) = \frac{S(U_t + h_r + h_1) - h_r U_t(T_f - T_a)}{h_2 U_t + h_2 h_r + h_2 h_1 + h_r U_t + h_r h_1} \quad (7.12.4)$$

$$(T_c - T_f) = \frac{h_r S - (h_2 + h_r)U_t(T_f - T_a)}{h_2 U_t + h_2 h_r + h_2 h_1 + h_r U_t + h_r h_1} \quad (7.12.5)$$

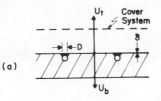
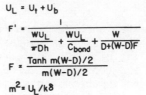

(a)

$$U_L = U_t + U_b$$

$$F' = \cfrac{1}{\cfrac{WU_L}{\pi Dh} + \cfrac{WU_L}{C_{bond}} + \cfrac{W}{D+(W-D)F}}$$

$$F = \frac{\text{Tanh}\, m(W-D)/2}{m(W-D)/2}$$

$$m^2 = U_L/k\delta$$

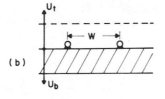
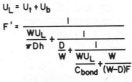

(b)

$$U_L = U_t + U_b$$

$$F' = \cfrac{1}{\cfrac{WU_L}{\pi Dh} + \cfrac{1}{\cfrac{D}{W} + \cfrac{1}{\cfrac{WU_L}{C_{bond}} + \cfrac{W}{(W-D)F}}}}$$

$$F = \text{same as (a)}$$

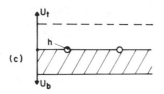

(c)

$$U_L = U_t + U_b$$

$$F' = \cfrac{1}{\cfrac{WU_L}{\pi Dh} + \cfrac{W}{D+(W-D)F}}$$

$$F = \text{same as (a)}$$

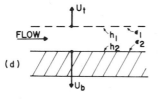
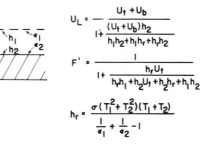

(d)

$$U_L = \cfrac{U_t + U_b}{1 + \cfrac{(U_t+U_b)h_2}{h_1 h_2 + h_1 h_r + h_r h_2}}$$

$$F' = \cfrac{1}{1 + \cfrac{h_r U_t}{h_1 h_1 + h_2 U_t + h_2 h_r + h_1 h_2}}$$

$$h_r = \frac{\sigma(T_1^2 + T_2^2)(T_1 + T_2)}{\cfrac{1}{\epsilon_1} + \cfrac{1}{\epsilon_2} - 1}$$

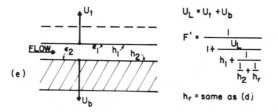

(e)

$$U_L = U_t + U_b$$

$$F' = \cfrac{1}{1 + \cfrac{U_L}{h_1 + \cfrac{1}{\cfrac{1}{h_2} + \cfrac{1}{h_r}}}}$$

$$h_r = \text{same as (d)}$$

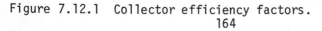

Figure 7.12.1 Collector efficiency factors.

164

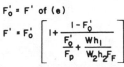

(f)

FLOW NORMAL TO PAPER

$F_o' = F'$ of (e)

$$F' = F_o' \left[1 + \frac{1 - F_o'}{\frac{F_o'}{F_p} + \frac{Wh_1}{W_2 h_2 F_F}} \right]$$

F_p = fin efficiency of plate

F_F = fin efficiency of fin

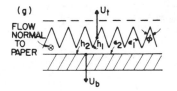

(g)

FLOW NORMAL TO PAPER

$U_L = U_t + U_b$

U_t is based on projected area

F' = same as (e) with h_1 replaced by $h_1 / \sin \phi/2$

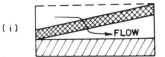

Clear glass   Black glass

FLOW

(h)

See Selcuk (1971)

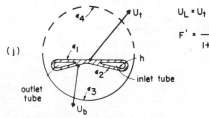

(i)

FLOW

See Hamid & Beckman (1971)

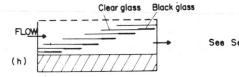

(j)

outlet tube

inlet tube

$U_L = U_t + U_b$

$$F' = \frac{1}{1 + \frac{U_L}{h}}$$

Valid only with neglible heat transfer between inlet and outlet fluid tubes.

Figure 7.12.1 (Continued)

165

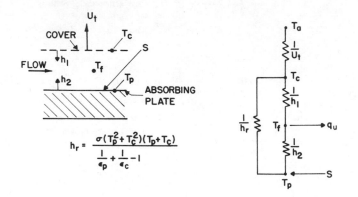

Figure 7.12.2 Type d solar air heater and thermal network.

Substituting these into the equation for q_u and rearranging, we obtain

$$q_u = F'[S - U_L(T_f - T_a)] \qquad (7.12.6)$$

where

$$F' = \left(1 + \frac{h_r U_t}{h_r h_1 + h_2 U_t + h_2 h_r + h_1 h_2}\right)^{-1} \qquad (7.12.7)$$

$$U_L = \frac{U_t}{1 + (U_t h_2)/(h_1 h_2 + h_1 h_r + h_2 h_r)} \qquad (7.12.8)$$

Note that U_L for this collector is not just the top loss coefficient in the absence of back losses but also accounts for heat transfer between the absorbing surface and the bottom of the cover. Whenever the heat removal fluid is in contact with a transparent cover, U_L will be modified in a similar fashion.

In order to account for back losses, a resistance could be added from the plate to the surroundings, and the same basic procedure could be followed. However, the algebra becomes extremely complicated and the resulting expressions for F' and U_L are unnecessarily cumbersome. Since the back losses in a well-designed system are always small, it is sufficient to add the back loss heat transfer coefficient, U_b, to the top loss coefficient, U_t. This assumes that back losses occur from T_c instead of T_b, but the resulting error should be small.

Example 7.12.1 Calculate the performance of a single-cover type *e* air heater of Figure 7.12.1 when the radiation incident on the collector is 900 W/m². The plate spacings are all 1 cm, the inlet temperature is 60 °C, the ambient air temperature is 0 °C, and the air mass flow rate is 200 kg/hr. The collector is 1 m wide by 4 m long. The emittances of the air duct surfaces are 0.95. The effective transmittance-absorptance product is 0.88. The wind speed is 5 m/s.

From Figure 7.4.4*b* with an assumed average plate temperauure of 70 °C, the top loss coefficient is 7.2 W/m² °C.

The radiation coefficient between the two air duct surfaces is calculated by assuming a mean temperature for radiation of 340 K (this can be checked later but is not critical).

$$h_r = \frac{4\sigma\bar{T}^3}{(1/\varepsilon_1) + (1/\varepsilon_2) - 1} = \frac{4 \times 5.67 \times 10^{-8} \times 340^3}{(2/0.95) - 1}$$

$$= 8.1 \text{ W/m}^2 \text{ °C}$$

The heat transfer coefficients between the air and the duct walls will be assumed to be equal. The characteristic length is the hydraulic diameter, which for flat plates is twice the plate spacing. The Reynolds number is

$$Re_{D_H} = \frac{\rho V D_H}{\mu} = \frac{\dot{m} D_H}{A_f \mu} = \frac{200 \times (2 \times 0.01)}{(0.01 \times 1) \times 0.065} = 6150$$

The length to diameter ratio is

$$\frac{L}{D_H} = \frac{4}{2 \times 0.01} = 200$$

Since Re is greater than 2100 and L/D_H is large, the flow is turbulent and fully developed and from Eq. (4.13.1)

$$Nu = 0.0158(6150)^{0.8} = 17$$

The heat transfer coefficient is then

$$h = 17 \times \frac{k}{D_H} = \frac{17 \times 0.029}{2 \times 0.01} = 25 \text{ W/m}^2 \text{ °C}$$

From Figure 7.12.1*e*

$$F' = \left(1 + \frac{U_L}{h + [(1/h) + (1/h_r)]^{-1}}\right)^{-1} = 0.81$$

From Eq. (7.7.4),

$$F'' = \frac{F_R}{F'} = \frac{(200/4) \times 1.18}{3.6 \times 7.2 \times 0.81}\left[1 - \exp\left(-\frac{3.6 \times 7.2 \times 0.81}{(200/4) \times 1.18}\right)\right] = 0.84$$

or

$$F_R = F''F' = 0.68$$

The useful gain is then

$$Q_u = 4 \times 0.68[900 \times 0.88 - 7.2(60 - 0)] = 979 \text{ W}$$

and the efficiency is

$$\eta = \frac{Q_u}{4 \times 900} = \frac{979}{3600} = 0.27$$

and the outlet temperature is

$$T_{c,o} = T_{c,i} + \frac{Q_u}{\dot{m}C_p} = 60 + \frac{979 \times 3600}{200 \times 1000 \times 1.18} = 75 \text{ °C}$$

7.13   SHORT-TERM COLLECTOR PERFORMANCE

This chapter has been concerned with predicting instantaneous collector performance using Eq. (7.7.5). We have shown that basically three parameters dictate collector performance: F_R, U_L, and $\tau\alpha$. F_R is a weak function of temperature through the temperature dependence of U_L but can usually be considered as constant for a given design. When F_R is on the order of 0.95, increasing U_L by a factor of 2 will decrease F_R by only 0.04. This same small effect would occur if the flow rate through the collector were doubled. Thus, wide variations in operating conditions produce only a small effect on F_R for most cases.

The overall loss coefficient U_L is a function of temperature. However, for many applications it is possible to select a single value of U_L. If this is not satisfactory, U_L can be evaluated as a function of temperature using, for example, Eq. (7.4.8).

Instantaneous collector efficiency is given on Figures
7.13.1a and 7.13.1b. For these figures, F_R and F' are both
specified so that it is not necessary to specify the mass flow
rate through the collector. The only difference between these
two figures is the level of the incident solar radiation. For
both of these figures, the value of U_L was calculated by solving the system of algebraic equations as was done in Example
7.4.1, except that back losses have been neglected.

If a constant value of U_L had been used to produce these
figures, all the curves would have been straight lines. For
example, in Figure 7.13.1a, if an inlet temperature of 100 °C
had been used to evaluate U_L for the two-cover, nonselective
collector, the performance curve would have been a straight
line from its value at 10 °C (η = 0.72) to its value at 100 °C
(η = 0.37). This approximation is reasonably accurate for inlet
temperatures not too far from 100 °C. If this same value of U_L
is used when the radiation level is reduced to 750 W/m^2, the
performance characteristics will appear as a straight line in
Figure 7.13.1b from the same point at 10 °C to the intersection
with 100 °C inlet temperature at an efficiency of about 25%.
Again, the straight line representation is not a bad approximation to the actual operating characteristics. Note that it is
usually better to make estimates of U_L based on days of good
radiation since these are the days that produce most of the
useful gain.

Finally, the transmittance-absorptance product can also be
considered as independent of direction, and therefore a constant, for many applications. The angular dependence of transmission and absorption have been shown to be small for incident
angles less than 45°. During the middle part of the day when
most collection occurs, the incident angle does not change significantly. If the orientation is such that the incident angle
is less than 45° during this period, a single value of $\tau\alpha$ can
be used. Even if the incident angle is greater than 45°, a
single value of $\tau\alpha$ may be adequate during the middle part of
the day if an appropriate average is used. The corrections to
$\tau\alpha$ to account for multiple reflections and absorption in the
glass [i.e., $(\tau\alpha)$ and $(\tau\alpha)_e$] are usually small for a well-designed collector.

The reason for the preceding discussion was to point out
that with well-chosen values of F_R, U_L, and $\tau\alpha$, collector performance can be predicted to a reasonable degree of accuracy.
These simplifications are vital when large-scale transient simulations are being considered in order to reduce computation
costs. As will be seen, a significant amount of system design
and analysis can be performed with the simple collector model.
These simplifications will be made in much of the remainder of
this book. It is often unnecessary to use the more complicated

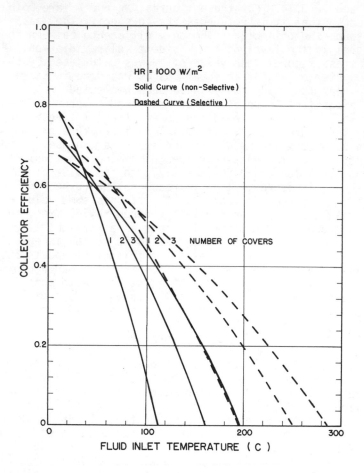

Figure 7.13.1a Collector efficiency as a function of fluid inlet temperature with HR = 1000 W/m². The following conditions
apply: F' = 0.95, F_R = 0.90, tilt = 45°, wind = 5 m/s, T_{amb}
= 10 °C, T_{sky} = 10 °C, ε_p = 0.95 (nonselective), ε_p = 0.10 (selective), KL= 0.0, $(\tau\alpha)$ = 0.87 (one cover), $(\tau\alpha)$ = 0.80 (two
covers), $(\tau\alpha)$ = 0.75 (three covers).

170

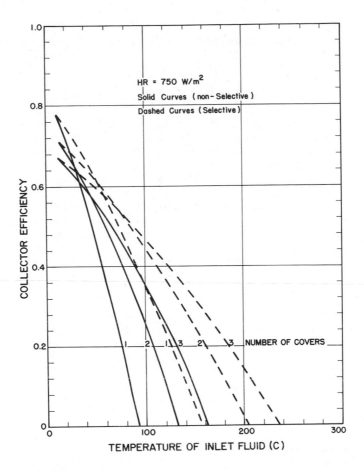

Figure 7.13.1*b* Collector efficiency as a function of fluid inlet temperature with HR = 750 W/m². The following conditions
apply: F' = 0.95, F_R = 0.90, tilt = 45°, wind = 5 m/s, T_{amb}
= 10 °C, T_{sky} = 10 °C, ε_p = 0.95 (nonselective), ε_p = 0.10 (selective), KL = 0.0, $(\tau\alpha)$ = 0.87 (one cover), $(\tau\alpha)$ = 0.80 (two
covers), $(\tau\alpha)$ = 0.75 (three covers).

171

time and temperature dependent functions for the three collector parameters.

7.14   LONG-TERM COLLECTOR PERFORMANCE

Thus far, we have examined the instantaneous performance of a
solar collector.   In later chapters, we will integrate solar
collectors into systems and evaluate system performance.   However, under some circumstances it is possible to separate the
collector from the system and evaluate long-term collector performance.   Hottel and Whillier (1958) developed a method of
combining the solar weather with collector performance into a
single relationship they call "utilizability."   Liu and Jordan
(1963) have extended the concept and introduced "generalized
utilizability curves."   However, there are significant differences in predicted performance between using the generalized
curves and using the curves for a specific location.   The reason
for these differences is not clear and further study is necessary.

It is apparent from Eq. (7.7.5) that for a given collector
at a specified tilt and entering temperature above ambient,
there is a value of the solar radiation at which the absorbed
solar energy and the thermal losses are equal.   This minimum
critical value of horizontal solar radiation is given by

$$H_c = \frac{U_L \left(T_{f,i} - T_a \right)}{R(\tau\alpha)}$$
(7.14.1)

The instantaneous useful energy gain of the collector can now
be expressed in terms of H_c, when $(H - H_c)$ is positive, as

$$Q_u = A_c F_R R(\tau\alpha)(H - H_c)$$
(7.14.2)

In this expression, F_R is a constant but R and $(\tau\alpha)$ vary with
the time of day.   However, for any particular hour pair, before
and after solar noon, $R(\tau\alpha)$ is nearly constant for an entire
month.   Consequently, it is possible to integrate Eq. (7.14.2)
for each pair of hours using measured data for H.

The average useful energy gain for a particular hour from
noon is found from

$$\frac{Q_u}{A_c} = F_R R(\tau\alpha) H_{\text{ave}} \left(\frac{1}{n}\right) \sum_n \left(\frac{H}{H_{\text{ave}}} - \frac{H_c}{H_{\text{ave}}}\right)^+$$
(7.14.3)

where n is the number of hours used in the calculation. The superscript + indicates that only positive values are to be used in the summation. Actually, many years of data for a particular month should be used to determine this average.

Hottel and Whillier define the utilizability, ϕ, as

$$\phi = \left(\frac{1}{n}\right) \sum_n \left(\frac{H}{H_{ave}} - \frac{H_c}{H_{ave}}\right)^+ \qquad (7.14.4)$$

When ϕ-curves were plotted as a function of H_c/H_{ave} for one location but for different hour pairs, it was found that there was little difference between the curves for the various hours. Consequently, the curves constructed for one hour pair, say 10:00 to 11:00 and 13:00 to 14:00, suffice for the whole day. Figure 7.14.1 gives a ϕ-curve for Blue Hill, Massachusetts, which was constructed using data from 1949 through 1951. It should be noted that the ϕ-curves are similar throughout the day but that H_{ave} varies hour by hour. Values of H_{ave} for various hour pairs for eight different months are also given on Figure 7.14.1 for Blue Hill.

In order to use ϕ-curves to predict long-term performance, it is necessary to know long-term averages of the critical radiation level, H_c. If the monthly average difference between the ambient temperature and the inlet fluid temperature is known for each hour, then the average value of H_c can be calculated. (As will become apparent when complete systems are considered, many systems operate with wide variations of the collector inlet temperature. Under these circumstances, it is difficult to determine a monthly average of H_c for each hour.)

> *Example 7.14.1* Calculate the long-term average January performance for the hour pair 10:00 to 11:00 and 13:00 to 14:00 of a collector operating in Blue Hill. Assume the collector has an overall loss coefficient of 8.0 W/m^2 °C, an effective transmittance-absorptance product of 0.88, an F_R of 0.9, and dirt and shading factors of 2% and 3%. Assume the average temperature difference between the collector inlet temperature and ambient temperature is 60 °C. Assume the collector is tilted toward the equator at an angle equal to the latitude of Blue Hill (42°).
>
> The average value for the declination for January from Eq. (2.5.1) is
>
> $$\delta = 23.45 \sin\left(360 \frac{284 + 15}{365}\right) = -21.3$$

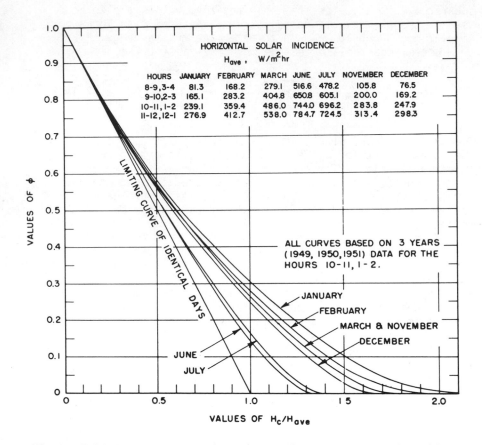

Figure 7.14.1 φ-curves for Blue Hill [from Whillier (1953)].

From Eq. (3.6.2), with ω = -22.5,

$$R = \frac{\cos\ (-21.3)\ \cos\ (-22.5)}{\cos\ (42)\ \cos\ (-21.3)\ \cos\ (-22.5) + \sin\ (42)\ \sin\ (-21.3)}$$

$$= 2.17$$

From Figure 7.14.1

$$H_{ave} = 239\ W/m^2$$

$$H_c = \frac{8 \times 60}{2.17 \times 0.88} = 251$$

$$\frac{H_c}{H_{\text{ave}}} = \frac{251}{239} = 1.05$$

Also from Figure 7.14.1, $\phi = 0.28$.
Finally, from Eq. (7.14.3),

$$\frac{Q_u}{A} = F_R R(\tau\alpha) H_{\text{ave}}(1 - d)(1 - s)\phi$$

$$= 0.9 \times 2.17 \times 0.88 \times 239 \times 0.98 \times 0.97 \times 0.28 = 109 \text{ W/m}^2$$

The average collector efficiency for this hour is

$$\eta = \frac{Q_u/A}{H_{\text{ave}}R} = F_R(\tau\alpha)\phi(1 - d)(1 - s) = 0.21$$

This same procedure could be followed to find the average performance for each hour pair, from which the average daily performance could be calculated.

7.15 *PRACTICAL CONSIDERATIONS FOR FLAT-PLATE COLLECTORS*

In the preceding sections of this chapter we have discussed the thermal performance of flat-plate collectors. There are also many practical considerations in the manufacturing, shipping, installation, and long-term use of flat-plate collectors.

In the design and manufacturing of collectors, primary consideration must be given to the ultimate cost of delivered energy. The addition of an extra cover glass may increase the thermal performance but it will also increase the cost of the collector. Thus the fundamental design choices, such as tube in sheet, evacuated glass tubes, and so on, can only be made when due consideration is given to the ultimate cost of delivered energy. The economics of specific applications will be more fully discussed in later chapters.

The practical problem of how the cover glasses should be mounted to prevent thermal stress breakage must be considered. The magnitude of this problem becomes apparent when it is realized that some collectors can reach temperatures of 200 °C

above the ambient if the heat removal fluid is not being circulated. The system must be designed to survive at these elevated temperatures because at some time the pumping system will
fail. In fact, during installation the circulating pumps may
not be connected.

Consideration must also be given to low-temperature operation. If the system operates in temperate climates, a freezing
problem exists for water heaters and the system must either be
drained or anti-freeze must be added.

Other types of weather problems exist. The system must be
able to withstand wind loads, hail, and rain. The collectors
must be designed so that snow does not interfere with winter-
time operation.

At some time failures will occur, and accessability and
maintenance become important. A repairman must be able to replace or repair equipment without excessive expense to the
owner.

The shipping and installation of collectors present problems. The size of the finished package, if it is a complete
factory-built unit, must be such that it can be shipped without
damage and installed by two men on the job site. If the units
are shipped in pieces, the training of the installers must be
considered.

However, many thousands of flat-plate solar-water heaters
have been sold commercially.

REFERENCES

Bliss, R. W., Solar Energy, *3*, 55, No. 4 (1959). "The Deriva-
tions of Several 'Plate Efficiency Factors' Useful in the Design
of Flat-Plate Solar-Heat Collectors."

Chiou, J. P., El-Wakil, M. M., and Duffie, J. A., Solar Energy,
9, 73 (1965). "A Slit-and-Expanded Aluminum-Foil Matrix Solar
Collector."

Dietz, A. G. H., Introduction to the Utilization of Solar Energy,
Zarem, A. M. and Erway, D. D., eds., New York, McGraw-Hill,
1963. "Diathermanous Materials and Properties of Materials."

Garg, H. P., Solar Energy, *15*, 299 (1974). "Effect of Dirt on
Transparent Covers in Flat-Plate Solar Energy Collectors."

Hamid, Y. H. and Beckman, W. A., Trans. ASME J. Engr. Power, *93*,
221 (1971). "Performance of Air-Cooled Radiatively Heated
Screen Matrices."

Hottel, H. C. and Woertz, B. B., Trans. ASME *64*, 91 (1942). "Performance of Flat-Plate Solar-Heat Collectors."

Hottel, H. C. and Whillier, A., Transactions of the Conference on the Use of Solar Energy, *2*, Part I, 74, University of Arizona Press, 1958. "Evaluation of Flat-Plate Collector Performance."

Klein, S. A., M. S. Thesis, University of Wisconsin, 1973. "The Effects of Thermal Capacitance Upon the Performance of Flat-Plate Solar Collectors."

Klein, S. A., Duffie, J. A., and Beckman, W. A., ASME, J. Engr. Power, *96A*, 109 (1974). "Transient Considerations of Flat-Plate Solar Collectors."

Liu, B. Y. H. and Jordan, R. C., Solar Energy, *7*, 53 (1963). "A Rational Procedure for Predicting the Long-Term Average Performance of Flat-Plate Solar-Energy Collectors."

Selcuk, K., Solar Energy, *13*, 165 (1971). "Thermal and Economic Analysis of the Overlapped-Glass Plate Solar-Air Heaters."

Speyer, E., Trans. of ASME, J. Engr. Power, *86*, 270 (1965). "Solar-Energy Collection with Evacuated Tubes."

Tabor, H., Bull. Res. Coun. Israel, *6C*, 155 (1958). "Radiation, Convection, and Conduction Coefficients in Solar Collectors."

Whillier, A. and Saluja, G., Solar Energy, *9*, 21 (1965). "Effects of Materials and of Construction Details on the Thermal Performance of Solar Water Heaters."

Whillier, A., Low Temperature Engineering Applications of Solar Energy, New York, ASHRAE, 1967. "Design Factors Influencing Collector Performance."

Whillier, A., ScD. Thesis, MIT, 1953. "Solar Energy Collection and Its Utilization for House Heating."

8. FOCUSING COLLECTORS

Focusing collectors utilize optical systems--reflectors or re-fractors--to increase the intensity of solar radiation on the energy-absorbing surface. Higher energy flux on that surface means a smaller surface area for a given total amount of ener-gy, and correspondingly reduced thermal losses. Consideration of the energy balances, which are basically similar to those for flat-plate collectors, shows that operation at higher tem-peratures is possible. While thermal losses are reduced, two additional kinds of losses become significant: most focusing systems operate only on the beam component of solar radiation, and the diffuse is lost; and additional optical loss terms be-come significant.

A focusing collector can be viewed as a special case of the flat-plate collector, modified by interposition of a radia-tion concentrator which serves to raise the otherwise low level of radiation on the absorber. Although the basic concepts of Chapter 7 are applicable to focusing systems, it is necessary to consider a number of complications arising from the use of the concentrator. Among the complications are: optical char-acteristics of concentrators, nonuniform fluxes on absorbers; wide variations in shape, temperature and thermal loss charac-teristics of absorbers, and the introduction of additional op-tical factors into the energy balance.

Focusing collectors can have radiation intensities at ab-sorbing surfaces increased by a wide range of ratios, from low values of 1.5 to 2, to high values of the order of 10,000. In-creasing ratios means increasing temperatures at which energy is delivered but it also means increasing requirements of pre-cision of optical systems, thus increasing costs. The cost of delivered energy from a focusing collector system is thus a di-rect function of the temperature at which it is available. At the highest range of concentration (and correspondingly highest precision of optics) focusing collectors are termed *solar fur-naces* and are laboratory tools for study of properties of ma-terials at high temperatures and similar purposes. Laszlo (1956) and the proceedings of a Solar Furnace Symposium (1957) include extensive discussions of solar furnaces. Our concern in this chapter is with energy delivery systems operating at low or intermediate concentrations.

From an engineering point of view, focusing collectors present some additional problems. They must (except at the very low end of the concentration scale) be oriented in varying degrees, to "track" the sun so that beam radiation will be directed onto the absorbing surface. However, the designer has open to him a range of configurations of the system which allow new sets of design parameters to be manipulated. There are also new requirements for maintenance, particularly to retain the quality of optical systems for long periods of time against dirt, weather, oxidation, and so on. The result of the combination of operating problems and collector cost has restricted the utility of focusing collectors, and no long-time practical applications of focusing collectors other than for furnaces are being made. New materials and better engineering of systems may make them of practical importance.

To avoid confusion of terminology, the word *collector* will be applied to the total system including the receiver and the concentrator. The *receiver* is that element of the system where the radiation is absorbed and converted to some other energy form, and includes the *absorber* and associated covers, insulation, and so on. The *concentrator*, or *optical system* is the part of the collector that directs radiation onto the receiver.

In this chapter, we write the governing equations in general terms, but the large number of possible collector configurations makes it difficult to develop generally useful relationships for thermal losses and other terms in the energy balances. Solutions for particular configurations are shown as examples. In general the engineer must apply good optics, heat transfer, thermodynamics, and design practice to the particular configurations he selects for evaluation.

The chapter is introduced by a brief discussion of theoretical solar images. It is shown that solar images from practical concentrators can be much larger than the theoretical. Then, after mentioning some concentrator-receiver configurations, energy balance concepts are discussed. Not much information has been published concerning energy balance measurements on focusing collectors; however, a study has been made of a cylindrical parabolic reflector system, and the data are used to illustrate the performance of the system and a method of calculating system performance. There are other methods, which are not discussed here, for analyzing and describing the optical properties of concentrators; ray-tracing techniques and general computer programs for evaluating optical systems constitute useful tools for more detailed evaluation of the optical problems of focusing systems.

8.1 *THE SOLAR DISK AND THEORETICAL SOLAR IMAGES*

It is the function of an optical system to form an image of the
sun on a receiver, the image being, in general, not a distinct
one. The relationships of the sun's size and distance from
earth determine that the angle subtended by the solar disk is
32' to an observer on earth. Thus, a theoretical image of the
sun created by any optical system will have a finite size, which
is dependent on the size of the solar disk and on the geometry
of the system. This is illustrated in Figure 8.1.1, where W'
(or W) is the theoretical image size (width or radius) for an
image formed by any part of the reflector. For a receiver that
is planar and normal to the axis of the concentrator, the width,
W', is found using the applicable "mirror radius," r, from

$$W' = \frac{2r \tan 16'}{\cos \phi} \qquad (8.1.1)$$

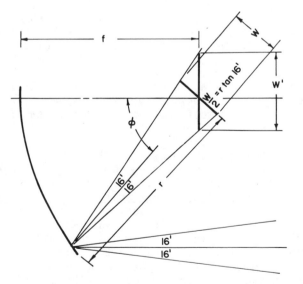

Figure 8.1.1 Schematic of theoretical solar image formed by a
concentrator. The ray shown is very nearly incident on the
edge of the reflector; at the edge ϕ is ϕ_{max}, the rim angle,
and r is r_{max}.

The distance r from a point on the reflector to the focus can
be derived for the particular reflector shape. For a reflector
of parabolic section, the focal length is defined by the equation of the surface

$$y^2 = 4fx \qquad\qquad (8.1.2)$$

and r is given by

$$r = \frac{2f}{1 + \cos \phi} \qquad\qquad (8.1.3)$$

where f is the focal length of the parabola, and ϕ is the angle
between the axis and the reflected beam at the focus as shown
in Figure 8.1.1. Note that as ϕ varies from zero to ϕ_{max}, r
increases from f to r_{max} and the theoretical image size in-
creases from $W|_{r\,=\,f}$ to $W'|_{r\,=\,r_{max}}$ (i.e., from $2f$ tan 16' to
$2r_{max}$ tan 16'/cos ϕ_{max}). Thus, there is a finite image size
and spreading of the image even for geometrically perfect systems. Figure 8.1.2 shows a section of an ideal solar image on
a plane normal to the axis of a parabola.

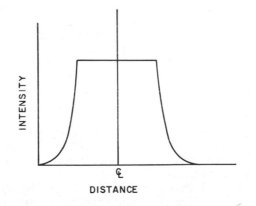

Figure 8.1.2 Cross section of a theoretical solar image on a
surface normal to the axis of a parabolic reflector, assuming
a uniform solar disk.

The *aperture* is the opening or projected area of the optical system. For surfaces of revolution, the aperture is usually characterized by the diameter of the reflector, or for

cylindrical systems, by the width. The focal length is a determining factor in image size as shown by Eqs. (8.1.1) and
(8.1.3), and the aperture, a, the determining factor in total
energy; thus the image brightness or energy flux concentration
at the focus of a focusing system will be a function of the
ratio a/f.

Radiation is not received uniformly from the total solar
disk. As noted by de la Rue et al. (1957), the center of the
sun is brightest, with the limbs (i.e., the edge of the sun)
darkest. Thus there is an additional nonuniformity of the distribution of radiation across a theoretical solar image that
would tend to distort the theoretical image of Figure 8.1.2.

This section has referred to theoretical images and applies
only to very precise optical systems. Most solar reflectors of
potential practical interest are not precise optical instruments, and produce images substantially larger than the theoretical; it is with these larger images that we will be concerned.

8.2 CONCENTRATORS, RECEIVERS, AND ORIENTING SYSTEMS

There is a wide variety of means for increasing the flux of
radiation on receivers; they can be classified as lenses or reflectors by the types of mounting and orienting systems, by the
concentration of radiation they are able to accomplish, by materials of construction, or by application. A characteristic
of primary importance is the concentration ratio, A_a/A_r, the
ratio of area of the concentrator aperture to the energy absorbing area of the receiver.

Figure 8.2.1a shows sections of several types of focusing
reflector systems. Any of the systems shown can use either cylindrical reflectors or refractors, focusing radiation more or
less sharply on a "line," or circular reflectors or refractors
which focus radiation on a "point" receiver. Concentration ratios for surface of revolution reflectors can be much higher
than their cylindrical counterparts. The following comments are
predicated largely on circular concentrators, but the same general comments are true for either.

Figure 8.2.1a shows a flat receiver with plane reflectors
at the edges to reflect additional radiation onto the receiver.
The concentration ratios of this type are relatively low, with
a maximum value of less than four. Some of the diffuse component of radiation incident on the reflectors would be absorbed
at the receiver; an analysis of a vee-trough collector of this
type has been presented by Hollands (1971). Figure 8.2.1.b
shows a next step, a conical system in which the receiver is

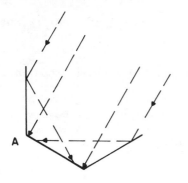
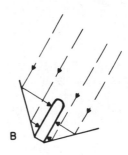
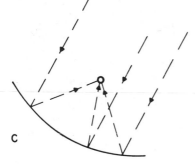
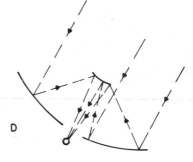
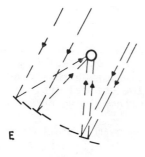
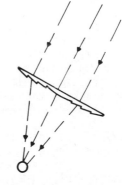

Figure 8.2.1 Some possible focusing system configurations:
(a) Plane receiver, plane reflector, (b) Conical reflector,
cylindrical receiver, (c) Paraboloidal concentrator, (d) Paraboloidal concentrator with secondary reflector, (e) Fresnel
reflector, (f) Fresnel refractor.

cylindrical. Cobble (1963) has analyzed the optical properties
of these systems.

Figure 8.2.1*c* shows a parabolic system that lends itself
to very high concentration ratios, which could in principle be
used in applications where high temperatures are required. A
modification is shown in Figure 8.2.1*d*, where an auxiliary or
secondary reflector is used in effect to shift the focus to a
more convenient location. Other arrangements of secondary reflectors are possible.

A *Fresnel reflector* is shown in Figure 8.2.1*e* and its refracting counterpart in 8.2.1*f*. The individual plane or curved
segments are each designed to reflect radiation to the receiver.
The advantage of the system is in its lack of appreciable dimension in the direction normal to the radiation, which may permit
easy fabrication; a disadvantage lies in the lost areas between
the segments near the rim of the assembly.

There is considerable latitude in receiver shape and design
for focusing systems. Flat receivers may be used for any but
type *b* systems. Cylindrical, hemispherical, or other curved
shapes may be possible or appropriate for any of the systems
illustrated, and cavity receivers may be used with types *c* to *f*.
Principles of receiver design and performance will be discussed
in a later section.

Although the systems operating at very low concentration
ratios may focus a part of the diffuse radiation onto the receiver, in general, the optical (focusing) devices are effective
only on beam radiation. It is evident also that the orientation
of the concentrator and receiver relative to the direction of
propagation of the beam radiation is important, and that "sun
tracking," in some degree, will be required for focusing systems. A variety of orienting mechanisms have been designed to
move focusing exchangers so that the incident beam radiation
will be reflected to the receiver. The motions required to accomplish tracking vary with the design of the optical system,
and a particular resultant motion may be accomplished by more
than one system of component motions.

Cylindrical optical systems will focus beam radiation to
the receiver (neglecting end effects) if the focal axis, the
vertex line of the reflector, and the sun lie in a plane. Thus,
for this type of system, it is possible to rotate the optics
about a single axis to meet this requirement; this axis of rotation may be north-south, east-west, or inclined and parallel
to the earth's axis (in which case the rate of rotation is
15°/hr).

Reflectors that are surfaces of revolution generally must
be oriented so that the focus, vertex, and sun are in line, and

thus must be able to move about two axes. These axes may, for example, be horizontal and vertical. Or one axis of rotation may be inclined so that it is parallel to the earth's axis of rotation (i.e., a polar axis) and the other perpendicular to it. The effect of mode of orientation on the angle of incidence of beam radiation is summarized for seven cases in Table 8.2.1 from Eibling et al. (1953).

TABLE 8.2.1 Angles of Incidence of Beam Radiation on Exchangers with Various Orientations [From Eibling et al. (1953)].

Orientation of Collector	Incidence Factor, $\cos \theta_i =$
A. Fixed so that it is normal to the solar beam at noon on the equinoxes.	$\cos \delta \cos \omega$
B. Rotation about a horizontal, east-west axis with a single, daily adjustment permitted so that its surface-normal coincides with the solar beam at noon every day of the year.	$\sin^2 \delta + \cos^2 \delta \cos \omega$
C. Rotation about a horizontal east-west axis with continuous adjustment to obtain maximum energy incidence.	$(1 - \cos^2 \delta \sin^2 \omega)^{\frac{1}{2}}$
D. Rotation about a horizontal north-south axis with continuous adjustment to obtain maximum energy incidence.	$[(\sin \phi \sin \delta + \cos \phi \cos \delta \cos\omega)^2 + \cos^2 \delta \sin^2 \omega]^{\frac{1}{2}}$
E. Rotation about an axis parallel to the earth's axis with continuous adjustment to obtain maximum energy incidence.	$\cos \delta$
F. Rotation about two perpendicular axes with continuous adjustment to allow the surface normal to coincide with the solar beam at all times.	1

Further classification of orientation systems may be made on the basis of manual or mechanized operation. Manual systems depend on the observations of the operator and his skill at making the necessary corrections and may be adequate for some purposes if concentration ratio is not too high and if labor costs are not prohibitive. Manual systems are clearly out of consideration in industrialized economies.

Mechanized orienting systems can be divided into *sun-seeking systems* and *programmed systems*. Sun-seeking systems use detectors to determine system misalignment, and through control systems make the necessary corrections to realign the assembly. Programmed systems, on the other hand, cause the collector to be moved in a predetermined manner (e.g., 15°/hr about a polar axis as noted above) and may need only occasional checking to assure alignment. It may also be advantageous to use a combination of these tracking methods; for example, by superimposing small corrections by a sun-seeking mechanism on a programmed "rough-positioning" system. Any mechanized system must have the capability of adjusting the position of the collector from end-of-day position to that for operation early the next day.

The higher the temperature at which energy is to be delivered, the higher must be the concentration ratio and the more precise must be the optics of both the concentrator and the orientation system. As an illustration, Figure 8.2.2 shows the ranges of concentration ratios and types of optical systems needed to deliver energy at various temperatures.

8.3 *GENERAL CHARACTERISTICS OF FOCUSING COLLECTOR SYSTEMS*

As with flat-plate collectors, energy balances are used to describe the performance of focusing collector systems. In a manner analogous to flat-plate collectors, we will first describe the thermal performance in general terms and then in later sections we will describe the various terms in detail.

As an example, consider a parabolic cylindrical collector, as shown schematically in Figure 8.3.1. Per unit area of aperture, the energy balance at a location x may be written as

$$q_u = H_b R_b \rho \gamma \tau \alpha - U_L \frac{A_r}{A_a} (T_{r,x} - T_a) \qquad (8.3.1)$$

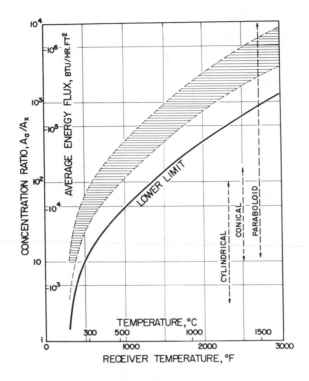

Figure 8.2.2 Relationships between concentration ratio and temperature of receiver operation. The "lower limit" curve represents concentration ratios at which the thermal losses will equal the absorbed energy; higher ratios will then result in useful gain. The shaded range corresponds to collection efficiencies of 40 to 60%, and represents the usual range of good operation. The flux density scale shows the average flux density of reflected radiation in the focal area. Also shown on the right side of the figure are approximate ranges in which several types of reflectors might be used.

(Note: This figure is not intended to be used for design. It is based on an assumed set of conditions determining the absorbed radiation and the thermal losses, and on good design practice at various temperatures. The positions of these curves would shift somewhat under conditions other than those assumed.) [From Duffie and Löf (1962).]

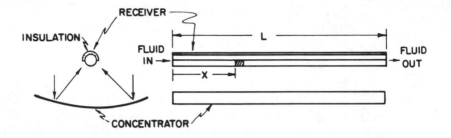

Figure 8.3.1 Schematic section and view of a cylindrical parabolic concentrator with cylindrical receiver shown insulated on
top portion.

or, if the whole receiver is at a uniform temperature, T_r (as
it might be if it were a boiler), the useful gain for the
collector is

$$Q_u = A_a H_b R_b \rho\gamma\tau\alpha - U_L A_r (T_r - T_a) \qquad (8.3.2)$$

where the various terms are as follows:

ρ. The specular reflectance of the reflector surface
(averaged over appropriate angles).

γ. The fraction of specularly reflected radiation that is
intercepted by the absorber surface is the intercept factor.

$\tau\alpha$. The transmittance of the cover (when a cover is present) and absorptance of the receiver.

$H_b R_b$. These terms have the same basic significance as for

flat-plate exchangers; H_b includes only the beam component of
the incident solar radiation (except for systems with very low
concentration ratio where some diffuse radiation may be collected); and R_b is the ratio of beam radiation on the reflector
aperture to that on whatever surface H_b may be measured. (For
beam radiation, some measurements on the surface normal to the
direction of propagation are available.) Thus the product
$H_b R_b$ is the beam radiation on the plane of the aperture of the
collector.

A_a/A_r. The ratio of effective area of the aperture to
area of the solar energy absorber is termed the *concentration
ratio*. Note that A_a is the unshaded projected area of the reflector system, which may be smaller than the total aperture
area.

For a flat-plate solar collector, $A_a = A_r$ and the maximum
energy flux of approximately 1 kW/m^2 limits the net useful heat
transfer rate in the exchanger. For focusing collectors, the
ratio A_a/A_r can range from unity to many thousands; in the
higher range, the energy flux rates can be comparable with those
in conventional industrial heat exchangers.

$U_L(T_r - T_a)$. The thermal losses per unit area from a receiver at a temperature T_r to the surroundings at T_a can often
be expressed by this linearized heat loss term. The evaluation
of this term is very similar to flat-plate collectors.

Since thermal losses from the large surface of a flat-plate
solar collector increase with temperature, low useful energy
delivery at high temperatures results. By increasing the concentration ratio, A_a/A_r, at constant A_a and U_L, thermal losses
are reduced. (This is equivalent to reducing U_L for a flat-
plate collector.) This method of controlling thermal losses
permits collection at higher temperatures. However, reflection
losses, intercept losses, and failure to collect diffuse radiation all reduce the absorbed energy from what it would be in a
flat-plate collector. The variation of useful heat recovery,
per unit area of collector aperture area A_a, as a function of
A_a/A_r, is shown in Figure 8.3.2 for an idealized example with
constant $H_b R_b \rho \gamma \tau \alpha$ product and receiver temperature. The useful
energy gain and efficiency curves approach asymptotes represent-
ing zero thermal losses at zero absorber area. In a practical
case, the absorbed energy also goes to zero at zero absorber
area. The balance between thermal losses and optical losses
will be considered in following sections.

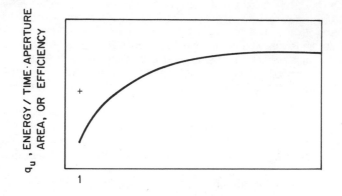

CONCENTRATION   RATIO

Figure 8.3.2 Variation of useful gain per unit area of aperture
with concentration ratio for an idealized example in which the
product $H_b R_b \rho \gamma \tau \alpha$ is taken as constant, and the thermal loss per
unit area of absorber surface is also constant. The point + at
A_a/A_r = 1 represents the performance of a flat plate collector
with the same $\tau\alpha$ and loss rate per unit area.

8.4 *OPTICAL LOSSES;* ρ *AND* $\tau\alpha$

The specular reflectance, ρ, is defined as the fractional portion of an incident collimated beam which is reflected such that
the angle of reflection equals the angle of incidence. It is a
function of the nature of the surface and of its smoothness.
High specular reflectance has commonly been achieved by use of
metal deposits or coatings on substrates of metals (front-surface reflectors) or glass (back-surface reflectors). Anodized
aluminum sheets have been used in experimental focusing units,
particularly of the parabolic cylinder type. Vacuum-metallized
plastic films, such as aluminized polyester film, have also been
used experimentally. Specular reflectances of a number of surfaces are given in Table 5.8.1. For practical focusing collector systems, it is important that high values of ρ be obtained
through the life of the collector.
    The transmittance, τ, and absorptance, α, have the same
significance as for flat-plate collectors. Their values may,
however, differ from those for flat-plate collectors, for the
following reasons which are dependent on the particular design
of a system: (a) frequently, focusing collectors do not have

covers, and τ does not enter the energy balance; (b) cavity absorbers are sometimes used as receivers which result in α approaching unity (with a corresponding high emittance); (c) τ and α are dependent on the average angle of incidence of the radiation on the cover and receiver. The angle of incidence of a beam of reflected radiation on the receiver will be a function of the position on the reflector from which the beam is reflected and the shape of the receiver. The proper value of the $\tau\alpha$ product must be arrived at by an integration of the radiation passing through the cover and incident on the receiver from all portions of the concentrator. By proper design, it should be possible to keep all angles of incidence less than 60° by shaping the receiver. These losses may be significant and their accurate analysis is necessary for proper description of exchanger performance; however, a rigorous analysis is very difficult, particularly for low-quality concentrators.

8.5 OPTICAL LOSSES; γ

The intercept factor represents the fraction of the specularly reflected radiation that is intercepted by the energy absorbing surface. This is an important concept as γ can, under some circumstances, represent a significant factor in the energy balance. The intercept factor is a property of the concentrator and its orientation in producing an image, and of the receiver and its positioning relative to the concentrator in intercepting part of that image.

Consider the flux distribution as a function of position in the focal area of a (cylindrical) collector, such as is shown in Figure 8.5.1.

The total area under the distribution curve is the total energy reflected to the focal plane. If a receiver occupies the width from A to B of this distribution, it will intercept energy represented by the shaded area. Thus the definition of γ, the fraction intercepted, can be written as

$$\gamma = \frac{\displaystyle\int_A^B I(w)\ dw}{\displaystyle\int_{-\infty}^{\infty} I(w)\ dw} \qquad\qquad (8.5.1)$$

where w is the distance from the center of the focal area. Similar considerations hold for concentrations that are surfaces of revolution.

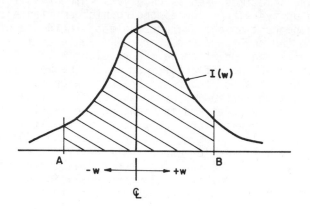

Figure 8.5.1 A flux distribution for a cylindrical reflector; the receiver size is *AB*.

The optimum performance of a system with a reflector of given optical properties will usually be obtained with the intercept factor being less than unity. In general, an optimum receiver size results in the maximum useful energy gain by minimizing the sum of optical and thermal losses; a large receiver results in large thermal and low optical loss while a small receiver means lower thermal loss but larger optical loss because of reduced intercept factor.

The distortion of the theoretical flux distribution as shown in Figure 8.1.2 to the distribution shown in Figure 8.5.1 can be viewed as originating from four more or less distinct causes:

1. Small-scale errors or irregularities in the reflector surface which cause dispersion of the image. This effect could be considered as diminishing the specular reflectance, ρ (and increasing the diffuse reflectance if the small errors are considered to arise from the nature of the

2. Macroscopic errors in the reflector, resulting in distortion of the solar image;

3. Errors in positioning of the receiver relative to the reflector; and

4. Errors in orienting the collector system, which result in image enlargement and displacement.

Each of these contributions to optical losses will be discussed below.

1. An angular dispersion, δ, of a surface can be defined as
the angular spread of the reflected beam from a perfectly collimated (or parallel) beam of radiation incident on the surface.
Dispersion is a function of the small-scale surface irregularities. The dispersion effect increases the size of the image at
the focus, and the increase in image size is proportional to
the mirror radius, φ, and focal length. The effect of dispersion angle can be considered as an addition to the solar intercept angle of 32', as shown in Figure 8.5.2. The image produced will be similar in shape to that shown in Figure 8.1.2,
but will be larger than the theoretical. Intercept factors for
cylindrical and circular concentrators have been determined as
a function of an angular dispersion by Löf and Duffie (1963).

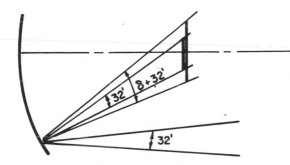

Figure 8.5.2 Schematic of a portion of a concentrator, with a
dispersion angle, δ, added to the 32' solar intercept angle.

2. The most significant errors causing enlargement of the
focal image in systems of low or intermediate concentration ratios (up to about 500) are macroscopic angular errors in concentrator construction. The nature of these errors, measured as
distortions or irregularities of the flux in the focal zone, is
dependent on the process of manufacture of the reflector, the
stiffness of its supporting structure, and other factors that
influence its shape. The nature of these errors and the magnitude of the effects on the energy balances are difficult to
predict, and few data are available on this subject.

Several methods have been proposed for describing this type
of error. Oman and Street (1960) use a mean angular deviation
of the reflector from parabolic shape. A related approach and
one somewhat easier to use experimentally is to describe the
reflector in terms of the distribution of the flux of radiation
in the image it produces. This procedure has been used by Liu
and Jordan (1965), and by Löf et al. (1962). The latter paper

is discussed below.

Assuming that the errors in reflector manufacture are random, a reasonable assumption for distribution or radiation across the focal area is a normal distribution. The solid curve of Figure 8.5.3 shows such a normal distribution curve for the flux in the focal area of a cylindrical parabolic reflector. The area under the curve is the total flux and is fixed by the product $H_b R_b A_a \rho \tau$; the sharpness of the distribution is then specified either by the standard deviation or the maximum flux density. The intercept factor can then be determined for a receiver of a given size by integrating across the distribution with limits fixed by the dimensions and position of the receiver relative to the distribution. The solid curve of Figure 8.5.4 shows γ versus receiver size for the normal distribution shown on Figure 8.5.3.

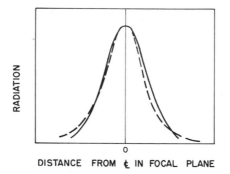

RADIATION

DISTANCE   FROM ₵ IN FOCAL PLANE

Figure 8.5.3 Normal distribution curve (solid-curve) compared to an experimental distribution curve for a parabolic cylindrical concentrator (dashed curve).

For cylindrical systems, the normal flux distribution can be written as

$$\frac{I}{I_{max}} = e^{-h^2 (w/W)^2} \tag{8.5.2}$$

and

$$I_{max} = \frac{1}{\sigma \sqrt{2\pi}} = \frac{h}{W \sqrt{\pi}} \tag{8.5.3}$$

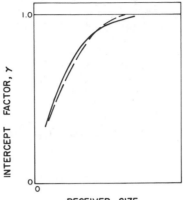

Figure 8.5.4 Intercept factor, γ, as a function of receiver size, for the two distributions shown in Figure 8.5.3. The solid curve is for the normal distribution; the dashed curve is for the experimental distribution.

where I = radiation flux density;
        I_{max} = the maximum flux, at the centerline;
        h = the normal flux distribution coefficient defined by Eq. 8.5.3;
        w = distance from the center of the zone;
        W = half-width of the concentrator; and
        σ = standard deviation of the normal distribution curve.

     For a symmetrical distribution, the intercept factor is given by

$$\gamma = \frac{2}{\sqrt{\pi}} \int_0^{h(w/W)_1} e^{-h^2(w/W)^2} \, d\left(\frac{hw}{W}\right) \qquad (8.5.4)$$

     Values of γ for this case are readily obtained by use of tables of the probability integral $(2/\sqrt{\pi})$ $\int_0^x e^{-x^2}dx$; entering the tables at a limit $x = h(w/W)_1$ gives values of γ for the particular width ratio and h. Curves such as the solid γ curve of Figure 8.5.4 can thus readily be constructed.

Example 8.5.1 What is the intercept factor for a parabolic cylinder concentrator producing an image in the focal plane with h = 60, if the receiver is symmetrical with respect to the center of the focus and the receiver has a width of 0.01 and 0.02 of the width of the concentrator?

Solution. For w/W equal to 0.01, the limit of integration of the probability integral is 60(0.01) = 0.60. From probability integral table, γ = 0.604. If the width is doubled to 1.20, γ = 0.910.

For a paraboloid, or more generally for a reflecting surface of revolution, the normal or Gaussian distribution function is

$$\frac{I}{I_{max}} = e^{-h^2(r/R)^2} \qquad (8.5.5)$$

and

$$I_{max} = \frac{1}{\sigma\sqrt{2\pi}} = \frac{h}{R\sqrt{\pi}} \qquad (8.5.6)$$

where I = radiative flux density at radial position from axis r/R;
  h = normal flux distribution coefficient, defined by Eq. (8.5.6);
  r = distance in focal plane from axis;
  R = radius of the concentrator; and
  σ = standard deviation of the normal distribution curve.

From the definition of intercept factor,

$$\gamma = \frac{I_{max}\int_0^{(r/R)_1} e^{-h^2(r/R)^2} \, 2\pi(r/R)d(r/R)}{I_{max}\int_0^{\infty} e^{-h^2(r/R)^2} \, 2\pi(r/R)d(r/R)} \qquad (8.5.7)$$

Integrating between limits,

$$\gamma = 1 - e^{-h^2(r/R)_1^2} \qquad (8.5.8)$$

This equation expresses the relationship between intercept factor and radius ratio and allows easy calculation of γ for systems with axial symmetry.

Example 8.5.2 What is the intercept factor for a paraboloidal concentrator and receiver with $(r/R)_1 = 0.02$ and 0.03, if $h = 60$, if the system is axially symmetric.

Solution. Using Eq. (8.5.8), for the first case

$$\gamma = 1 - e^{-(60)^2(0.02)^2} = 0.763$$

and for the larger receiver

$$\gamma = 1 - e^{-(60)^2(0.03)^2} = 0.961$$

Experimental measurement of flux distribution in the focal zone of a reflector may also be used to determine the intercept factor for that reflector as a function of receiver dimensions. It is necessary in these measurements to map the focal area with a radiation flux measuring device (such as thermocouple or photovoltaic detectors) taking enough traverses to establish the total flux pattern. An example of experimental measurements, by Löf et al. (1962), is shown by the dashed curve of Figure 8.5.3. The intercept factors resulting from this experimentally determined flux distribution for various receiver dimensions are shown on Figure 8.5.4. Figure 8.5.5 shows another experimental flux map obtained by photographic techniques for circular parabolic reflector.

Errors of reflector construction may not be random but may be systematic and reproducible results of the method of manufacture. Distribution functions other than the normal may be better under these circumstances.

3. Errors of positioning of the receiver relative to the reflector will result in image enlargement at the receiver in addition to that caused by reflector surface angular dispersion and gross reflector errors. This is illustrated in Figure 8.5.6, which indicates the enlargement of the focus and the reduced radiation intensity in the "focal area" for a parabolic reflector when the plane of the receiver is displaced from its proper position. The more a receiver of fixed area is displaced, the smaller is γ for a fixed receiver size.

4. Angular errors of focusing collector orientation, that is, pointing errors, result in both focal area enlargement and displacement. Thus the intercept factor will be reduced by such angular errors for a fixed receiver size due to spillage and lower flux density. For a given magnitude of the pointing error, the change of intercept factor will be a direct function of the focal length of the system and its concentration ratio.

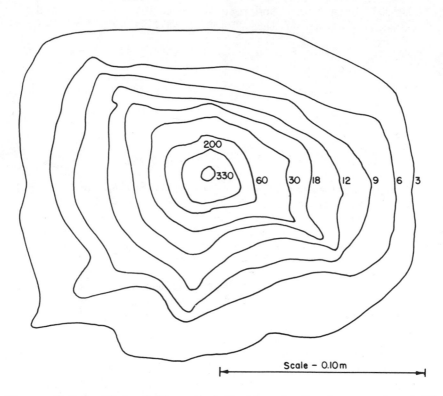

Figure 8.5.5 Plot of flux distribution in focal area of a 1.22
m diameter drape-molded polystyrene reflector. Lines are constant intensities, measured by photographic techniques, on a
plane normal to the axis of the paraboloid, in the apparent position of minimum focal area (i.e., at a distance from the vertex of 0.47 m for the nominal 0.46 m focal length reflector).
Numbers on the curve represent intensities relative to uncon-
centrated radiation intensity.

Experimentally, the effects of angular orientation errors or
errors of positioning of the receiver relative to the reflector
can be estimated from flux mapping procedures.
    Two additional factors to be considered are the rim angle
of the reflector and the shape of the receiver as they affect
γ and, thus, the performance of a system. The rim angle of the
reflector is defined as the angle between the axis and the line
drawn from the focus to the reflector rim; in Figure 8.1.1, the

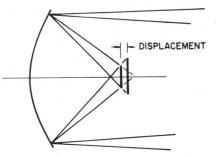

Figure 8.5.6 Effects of axial displacment of the receiver relative to the concentrator.

maximum value of ϕ is the rim angle. The rim angle is an important parameter describing, in effect, the limits of the directions from which the radiation reaches the receiver; the
shape of the receiver is thus dependent on it. The rim angle
fixes the size of the image on a receiver surface of a particular orientation and shape, thus fixing a minimum receiver size
to intercept the total solar image.
        Receivers may have a variety of shapes, depending on the
geometry of the concentrator; these can include flat, cylindrical, hemispherical, cavity, and other shapes. [An interesting
example of determination of receiver geometry for a circular-
cylindrical reflector, which led to a receiver of triangular
cross section, has been given by Tabor and Zeimer (1962).]
        Ray tracing methods can be used to show the effects of any
of these factors on the radiation flux distribution produced by
a concentrator and on the reflected radiation intercepted by a
receiver. Alternatively, experimental measurements on existing
collectors can provide this information for the concentrator on
which the measurements are made and for those of similar optical
characteristics.

        8.6 THERMAL PERFORMANCE OF FOCUSING COLLECTORS

The methods for calculating thermal losses from receivers of
focusing exchangers are not as easily summarized as in the case
of flat-plate exchangers, although the same principles are used.
For receivers, the shapes are widely variable, the temperatures
are higher, the edge effects are more significant, conduction
terms may be quite high, and the problems may be compounded by

the fact that the radiation flux on receivers is not uniform.
Thus, substantial temperature gradients may exist across the
energy-absorbing surfaces. It is not possible to present a
single general method of estimating thermal losses, and ultimately each receiver geometry must be analyzed as a special
case..
     The nature of the thermal losses for receivers of focusing-
type collectors are in general the same as for flat-plate exchangers, and the same basic methods are used to calculate them.
Receivers may have covers transparent to solar radiation; the
outward losses from the absorber by convection and radiation to
the atmosphere are correspondingly modified and equations similar to those of Chapter 7 can be used to estimate their magnitude. As with flat-plate systems, the losses can be estimated
as being independent of the intensity of incident radiation,
although this may not be strictly true, particularly if a transparent cover absorbs appreciable solar radiation. In any event,
an effective transmittance-absorptance product can also be defined for focusing systems. Furthermore, with focusing systems
the intensity of radiation at the receiver is generally such
that only cover materials with low absorptance for solar radiation can be used without thermal damage to the cover. Conduction losses occur through any insulation on the nonirradiated
portions of the receiver and through any supporting structure
for the receiver. For an example of calculation methods for a
particular set of geometries (i.e., uncovered cylinders and cylinders with cylindrical covers), see Tabor (1955).
     The generalized analysis of a focusing-collector system is
very similar to that of a flat-plate collector. It is necessary
to derive appropriate expressions for F', the collector efficiency factor, and U_L, the loss coefficient. With F' and U_L
known, the collector outlet temperature can be calculated from
an expression that is similar to Eq. (7.7.5).
     Consider a parabolic cylinder reflector and an uncovered
circular receiver. Assume that there are no temperature gradients around the receiver tube. The heat transfer coefficient
between the fluid and the tube is h_i, and the loss coefficient
from the outside of the tube is U_L. Note that U_L accounts for
radiation essentially with the sky due to the reflector. The
loss coefficient U_L is found from

$$U_L = h_{wind} + h_r \qquad (8.6.1)$$

The linearized radiation coefficient, h_r, is found from

$$h_r = 4\sigma\varepsilon_r\bar{T}^3 \qquad (8.6.2)$$

where $\bar{T}$ is the mean temperature for radiation and ε_r is the emittance of the receiver.* Since the heat flux in a concentrating system may be hign, the heat transfer resistance from the outer surface of the receiving tube to the fluid should include the tube wall. The overall heat transfer coefficient (based on the outside tube diameter) from the surroundings to the fluid is

$$U_o = \left(\frac{1}{U_L} + \frac{D_o}{h_i D_i} + \frac{D_o \ln D_o/D_i}{2k}\right)^{-1} \qquad (8.6.3)$$

where D_i and D_o are the inside and outside tube diameters, h_i is the heat transfer coefficient inside the tube, and k is the tube thermal conductivity.

Rewriting the energy balances of section 8.3., the useful energy gain q_u per unit of collector length L can be expressed in terms of the local receiver temperature, T_r, as

$$q_u = \frac{A_a}{L}H_b R_b \rho\alpha\gamma - \pi D_o U_L(T_r - T_a) \qquad (8.6.4)$$

and also in terms of the energy transfer to the fluid as

$$q_u = \frac{\pi D_o(T_r - T_f)}{(D_o)/(h_i D_i) + (D_o \ln D_o/D_i)/2k} \qquad (8.6.5)$$

If we eliminate T_r from Eqs. (8.6.4) and (8.6.5), we have

$$q_u = F'\frac{A_a}{L}\left[S - \frac{A_r}{A_a}U_L(T_f - T_a)\right] \qquad (8.6.6)$$

and

$$F' = \frac{U_o}{U_L} \qquad (8.6.7)$$

*If a single value of h_r is not acceptable due to large temperature gradients in the flow direction, the collector can be divided into two or more collectors, each with a constant value of h_r.

where

$$A_r = \pi D_o L; \quad S = H_b R_b \rho \gamma \alpha$$

The form of Eqs. (8.6.6) and (8.6.7) is identical to Eqs. (7.5.16) and (7.5.18). If the same procedure is followed as was used to derive Eq. (7.7.5), the following equation results:

$$Q_u = A_a F_R \left[S - \frac{A_r U_L}{A_a}(T_{f,i} - T_a) \right] \qquad (8.6.8)$$

with the collector flow factor F'' equal to

$$F'' = \frac{F_R}{F'} = \frac{\dot{m} C_p}{A_r U_L F'} \left[1 - e^{-A_r U_L F'/\dot{m} C_p} \right] \qquad (8.6.9)$$

The same analysis applies to a receiver that is covered, but it is necessary to include the effective transmittance-absorptance product in S and to properly evaluate U_L to account for the added heat transfer resistances.

If a receiver of the type discussed above serves as a boiler, F' is the same as is given by Eq. (8.6.7), but F_R is then identically equal to F' as there is no temperature gradient in the flow direction. If a part of the receiver serves as a boiler and other parts as fluid heaters, the two or three sections of the receiver must be treated separately.

Example 8.6.1 A cylindrical parabolic concentrator unit of width of 2.5 m and a length of 10 m has a reflective lining with a specular reflectance of 0.85. The receiver is a cylinder, painted flat black, and surrounded by a glass cylindrical envelope. The absorbing cylinder has a diameter of 6 cm and the transparent envelope has a diameter of 9 cm. Other optical properties of this system are estimated as

$$(\tau \alpha)_e = 0.77; \quad \gamma = 0.94$$

The collector is designed to heat a fluid entering the absorber at 200 °C, at a flow rate of 500 kg/hr. The fluid has $C_p = 1.26$ kJ/kg °C.

The appropriate heat transfer coefficients are estimated to be as follows:

From the fluid inside to the surroundings,

$$U_o = 6.0 \text{ W/m}^2 \text{ °C.}$$

From absorber outer surface to the surroundings,

$$U_L = 7.0 \text{ W/m}^2 \text{ °C.}$$

If the incident beam radiation on the aperture of the collector is 700 W/m² and the ambient temperature is 25 °C, calculate the useful gain, exit fluid temperature, and efficiency of collection of beam radiation.

Solution. The solution is based on the energy balance, Eq. (8.6.8). The area of the pipe is:

$$A_r = \pi DL = \pi \times 0.06 \times 10 = 1.88 \text{ m}^2$$

Taking into account shading of the central part of the collector by the receiver,

$$A_a = (2.5 - 0.09)10 = 24.1 \text{ m}^2$$

To calculate F_R, we first calculate F' for this situation from Eq. (8.6.7):

$$F' = \frac{6}{7} = 0.86$$

Then F_R from Eq. (8.6.8):

$$\frac{\dot{m}C_p}{A_r U_L F'} = \frac{500 \times 1.26}{1.88 \times 7.0 \times 3.6 \times 0.86} = 15.5$$

$$F_R = 15.5 \, F'(1 - e^{-1/15.5}) = 0.83$$

The absorbed solar energy is

$$S = 700 \times 0.85 \times 0.77 \times 0.94 = 430 \text{ W/m}^2$$

and the useful gain is

$$Q_u = 24.1 \times 0.83\left[430 - \frac{1.88 \times 7.0}{24.1}(200 - 25)\right] = 6690 \text{ W}$$

The exit fluid temperature is

$$t_{c,o} = t_{c,i} + \frac{Q_u}{\dot{m}C_p} = 200 + \frac{6690}{500 \times (1.26/3.6)} = 238 \ ^{\circ}C$$

Efficiency of use of beam radiation is

$$\eta = \frac{6690}{700 \times 25} = 0.38$$

Notes 1. The "optical efficiency," $\rho\gamma\tau\alpha = 0.85 \times 0.77 \times 0.94 = 0.62$.
2. Typical total radiation for this situation would be 850 W/m^2; the efficiency based on total radiation is then about 31%.

8.7 HEAT CAPACITY EFFECTS

Heat capacity effects for focusing collectors can be estimated in the same way as for flat-plate collectors. Concentrators will operate at or very near ambient temperatures. The equipment of concern is the receiver and other equipment that change temperature as collector operation begins or proceeds. Since the receiver is smaller in area than an equivalent flat-plate collector, its heat capacity per degree of temperature change may be significantly lower. However, the temperature excursions of a receiver may be much greater than those of a flat plate, and thus the heat capacity effect may be significant.

Klein (1973) outlined two considerations in heat capacity effects for flat-plate collectors; a "collector storage effect" resulting from energy required to heat the collector from its nighttime temperature to its final operating temperature of the day, and a "transient effect" of collector operation under varying meteorological conditions. Both of these should also be considered for focusing systems. The temperature increase from (nighttime) ambients to final operating temperatures are predictable and the energy necessary to accomplish that change can be estimated. The transient effect is more difficult to predict; an interruption in beam radiation by clouds will probably reduce the radiation input to zero, resulting in larger temperature changes than for flat-plate collectors.

8.8 EXPERIMENTAL PERFORMANCE OF A FOCUSING COLLECTOR

The performance of focusing collectors depends on a wide variety of factors. As has been noted, no simple method of calculation can be used for all systems. Thus, the problem reverts to treatment of special cases, for example, Löf et al. (1962) and Tabor (1958). Löf et al. treat a particular collector in detail and the results of this study provide a useful illustration of performance of a focusing collector; the balance of this section is based on that paper.

The collector consisted of a parabolic cylinder reflector of aperture, 1.89 m; length, 3.66 m; and focal length, 0.305 m, with bare tubular receivers of several sizes coated with a nonselective black paint having an absorptance of 0.95. The system was mounted so as to rotate on a polar axis at 15°/hr. It was operated over a range of temperatures from near ambient (by heating cool water flowing at a high rate) to about 180 °C (with the high-temperature operation using boiling water, water solutions, or other liquids).

The intercept factors for various receiver sizes were determined from multiple measurements of the flux distribution at the focal tube; these distributions were averaged and the resulting mean distribution is shown in Figure 8.8.1. The distribution is very similar to a normal distribution curve, but is displaced from the position of the theoretical focus. The intercept factor that result from this distribution, with the receiver tubes fixed at the theoretical focus by the mechanical design of the system, are shown in Figure 8.8.2.

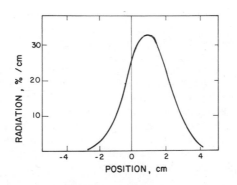

Figure 8.8.1 The experimental mean flux distribution for the parabolic cylinder reflector.

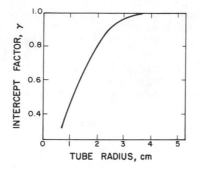

Figure 8.8.2   Intercept factors for tubes centered at position 0 of the reflector of Figure 8.8.1.

The results of many energy balance measurements are summarized in Figures 8.8.3 and 8.8.4, which show the distribution of incident beam solar energy (during operation at steady state in clear weather) into useful gain and various losses for two receiver tube sizes; the first for a 6 mm diameter tube and the second for a 2.7 mm diameter tube. The relative magnitudes of the losses are evident.

From these figures, it is possible to estimate the effects of design changes. For example, for this collector the use of selective surface of emittance 0.2 would reduce the radiation loss by 79% of the value shown at any temperature. However, radiation loss is not the dominant loss [a generalization made by Edwards and Nelson (1961)]. The most obvious initial improvements for this exchanger would be in reduction of optical losses by using surfaces of higher reflectance, and by intercept-factor improvements (the latter particularly for the smaller receiver).

In this study, the reflector and receiver tubes were supported by plates at each end; these provide a means of heat loss by conduction from the tubes. These losses were estimated from temperature measurements along the supporting plates. Although not shown in the figures, they were estimated at 3, 6, and 10% of the incident clear sky radiation for receiver-surface temperatures of 100, 135, and 175 °C, respectively, for the conditions of these experiments.

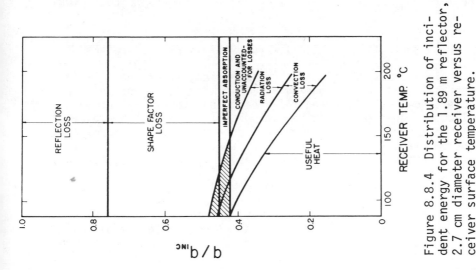

Figure 8.8.4 Distribution of incident energy for the 1.89 m reflector, 2.7 cm diameter receiver versus receiver surface temperature.

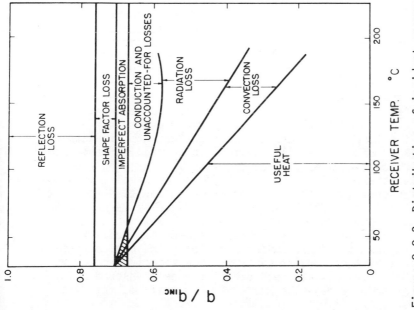

Figure 8.8.3 Distribution of incident energy for the 1.89 m reflector, 6 cm diameter receiver as a function of receiver surface temperature.

8.9 COLLECTOR OPTIMIZATION FOR MAXIMUM ENERGY DELIVERY

A complete study of exchanger optimization, to arrive at a design that will deliver energy at minimum cost, is not yet possible. However, the effects of design parameter changes on
collector performance can be evaluated and the optimum value of
that parameter selected on the basis of maximized performance
under a particular set of operating conditions. With a reflector of particular optical properties, a major design parameter
is the receiver size.

This question has been studied analytically by Liu and
Jordan (1965) for space application of paraboloids, and analytically and experimentally for the cylindrical system described
in the previous section by Löf et al. (1962). The results of
the Löf et al. study are summarized in Figure 8.9.1 which shows
efficiency at a particular set of conditions as a function of
receiver size (i.e., tube radius) for three absorber surface
temperatures. The tradeoffs between intercepting more radiation (i.e., higher γ) and increasing thermal losses are shown.
The dashed curves are experimental, from measurements of output and thus of efficiency. The solid curves are based on calculated energy balances, but using the appropriate experimental
values of the intercept factor, γ.

Collector geometries resulting in optimum thermal performance for systems which have negligible orienting error have
been calculated by Löf and Duffie (1963) as functions of the
product $H_b R_b \rho \tau \alpha$; a thermal loss rate per unit area of receiver,
q_{th}; and either of two parameters, h or δ, which describe the
optical precision of the reflector. The first of these parameters is h, the normal flux distribution coefficient for a normal
distribution curve as defined in section 8.5. The second is δ,
defined here as the angular error of curvature of a reflector from the theoretical contour, with uniform distribution of
angular error between the limits of $+\delta$ to $-\delta$, in degrees. Figure 8.9.2 shows examples of flux distributions arising from
each of these assumptions. Figure 8.9.3 shows sample computed
curves of useful gain per unit aperture area for two sets of
values of the parameters noted above; these curves are analogous
to the experimental curves shown in Figure 8.9.1.

The maxima of such curves have been plotted for a
range of values of each parameter describing the optical quality,
for cylindrical and circular systems. One set of these curves
is shown in Figure 8.9.4. Such curves can be developed using
other parameters to describe the optical quality of the system.

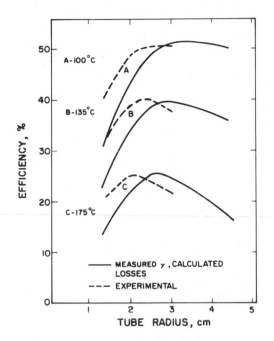

Figure 8.9.1 Efficiency as a function of receiver radius for the experimental cylindrical parabolic collector.

8.10 SPECIAL GEOMETRIES

Various combinations of reflector and receiver can be arranged to produce particular distributions of flux over the surface of the receiver. For example, for a conical reflector and cylindrical receiver shown in Figure 8.2.1*b*, the flux intensity at the receiver varies with the square of the distance from its lower end. By proper shaping of receivers with accurate parabolic reflectors, it is theoretically possible to get nearly uniform flux distribution on the receiver.

Another type of geometric system of possible interest is that of reflector-receiver geometries that require a minimum of tracking to compensate for the apparent motion of the sun. Tabor and Zeimer (1962) have presented an analysis of "stationary" reflectors, to be adjusted at weekly intervals to compensate for declination. The system was a circular cylindrical

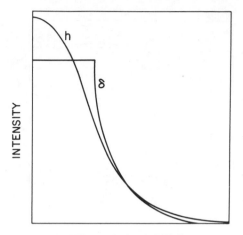

WIDTH or RADIUS RATIO

Figure 8.9.2 Examples of flux distributions resulting from
assumption of random angular reflector error between fixed limits (curve δ) and normal distribution (curve h). [From Löf and
Duffie (1963).]

reflector mounted on east-west axes, with receivers triangular
in cross section, and a concentration ratio of about three.
Systems of this type can also be developed that require occasional adjustments during the day; the upper limits of concentration ratio may then be of the order of perhaps 10 and a function of the frequency and precision of orientation. In these
systems, γ may vary substantially throughout the day or over
longer periods.

8.11 *MATERIALS AND CONSTRUCTION OF REFLECTORS*

Reflector construction entails some particular problems that
must be considered, some of which are listed below. For purposes of discussion, it is convenient to divide a reflector into two parts: (a) the reflective lining, and (b) the shell, supporting and orienting structure. The reflective lining and its
desirable characteristics were discussed in section 8.4. In
summary, it is desirable to use a reflective material with a
maximum specular reflectance over the period of use of the exchanger consistent with costs. It should also be pointed out

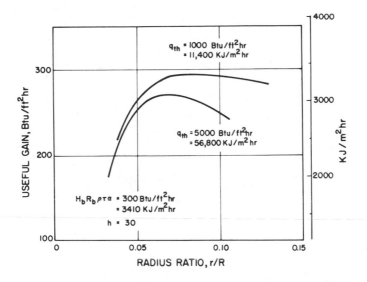

Figure 8.9.3 Effect of radius ratio on useful gain in Btu per square feet of projected reflector area at indicated rates of incident radiation and thermal losses. [From Löf and Duffie (1963).]

that there is the possibility of renewing a reflective lining, for example, by putting on a new layer of reflective plastic tape such as aluminized Mylar, or by replating or remetalizing, if the additional useful energy gain warrants the expenditure.

    The shell and supporting structure are of critical importance in their influence on γ, the intercept factor, and the operation of an exchanger is dependent on the ability of the structure to maintain the shape and orientation of the reflecting surface. Since orientation and shape are critical factors, a design must be made with the following conditions in mind:

    1. The shell and structure must be supported in various positions of orientation without significant distortion because of its own weight.
    2. They must be capable of being operated and therefore not significantly distorted in commonly encountered winds. Wind loading was shown by Baum (1956) to limit effectively the practical sizes of single concentrators.
    3. They must be capable of resisting structural damage in high winds and other storm conditions (e.g., hail) in a fixed

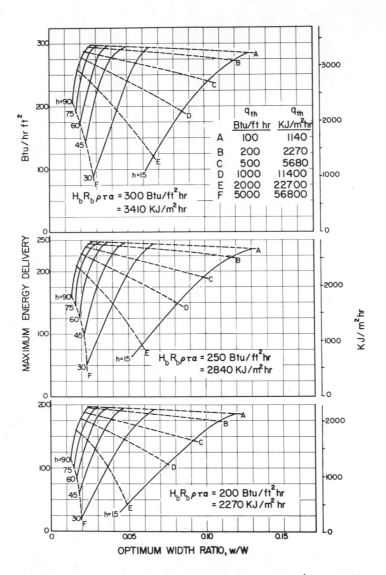

Figure 8.9.4 Maximum energy delivery, $q_u A_r/A_a$ (per unit area of reflector aperture) for various values of thermal loss rate, flux distribution coefficient h, and $H_b R_b \rho\tau\alpha$, for parabolic cylinder concentrators. q_{th} on the figure refers to the thermal loss rate. [From Löf and Duffie (1963).]

position, possibly secured.

In summary, this chapter points out the variety of simi-
larities and differences between flat-plate and focusing collectors. Perhaps the biggest difference lies in the fact that
all of the thousands of practical solar energy delivery devices
that have been manufactured, sold, and used in the world are
based on flat-plate concepts, and none on focusing systems (other
than for short-time or laboratory uses). Practical development
and application problems need to be solved before focusing collectors can become significant contributors to solar energy
processes. Research to solve these practical problems in focusing systems is being undertaken by various organizations (e.g.,
see the Colorado State University--Westinghouse Reports and the
University of Minnesota--Honeywell Reports to the National
Science Foundation).

REFERENCES

Baum, V. A., Proceedings of the World Symposium on Applied Solar
Energy, Menlo Park, Calif., Stanford Research Institute 289,
(1956). "Prospects for the Application of Solar Energy, and
Some Research Results in the USSR."

Cobble, M. H., Solar Energy, *7*, 75 (1963). "Analysis of a Conical Solar Concentrator."

Colorado State University--Westinghouse Report to the National
Science Foundation, Report NSF/RANN/SE/GI-37815/PR/73/3, (1973).
"Solar Thermal Electric Power Systems."

DelaRue, R., Lob, E., Brenner, J. L., and Hiester, N. K., Solar
Energy, *1*, No. 2-3, 94 (1957). "Flux Distribution Near the
Focal Plane."

Duffie, J. A. and Löf, G. O. G., Paper presented at World Power
Conference, Melbourne (1962). "Focusing Solar Collectors for
Power Generation."

Edwards, D. K. and Nelson, K. E., Paper 61-WA-158, presented at
1961 Winter ASME Meeting. "Radiation Characteristics in the
Optimization of Solar Heat-Power Conversion Systems."

Eibling, J. A., Thomas, R. E. and Laudry, B. A., Report to the
Office of Saline Water. U. S. Dept. of Interior (1953). "An
Investigation of Multiple-Effect Evaporation of Saline Waters
by Steam from Solar Radiation."

Hollands, K. G. T., Solar Energy, *13*, 149 (1971). "A Concentrator for Thin-Film Solar Cells."

Klein, S. A., M.S. Thesis in Chemical Engineering, Madison, University of Wisconsin (1973). "The Effects of Thermal Capacitance Upon the Performance of Flat-Plate Solar Collectors."

Laszlo, T. S., Image Furnace Techniques, New York, Wiley-Interscience, 1965.

Liu, B. Y. H. and Jordan, R. C., J. Engr. Power, Trans. ASME, *87*, 1-12 (1965). "Performance and Evaluation of Concentrating Solar Collectors for Power Generation."

Löf, G. O. G., Fester, D. A., and Duffie, J. A., J. Engr. Power, Trans. ASME, *84A*, 24 (1962). "Energy Balance on a Parabolic Cylinder Solar Reflector."

Löf, G. O. G.,and Duffie, J. A., J. Engr. Power, Trans. ASME, *85A*, 221,(1963). "Optimization of Focusing Solar-Collector Design."

Oman, H. and Street, G., Paper presented at AIEE meeting, San Diego, August 1960.

Proceedings of the 1957 Solar Furnace Symposium, Solar Energy, *1*, No. 2-3 (1957).

Tabor, H., Solar Energy, *2*, No. 1, 3 (1958). "Solar Energy Research: Program in the New Desert Research Institute in Beersheba."

Tabor, H., Bull. Res. Coun. Israel, *5C*, 5 (1955), "Solar Energy Collector Design." Also published in Trans. Conf. on Use of Solar Energy, II, Part I, 1, University of Arizona Press, 1958.

Tabor, H. and Zeimer, H., Solar Energy, *6*, 55, (1962). "Low Cost Focusing Collector for Solar Power Units."

University of Minnesota--Honeywell Report to the National Science Foundation, Report NSF/RANN/SE/GI-34871/PR/73/2 (1973). "Research Applied to Solar-Thermal Power System."

9. ENERGY STORAGE

Solar energy is a time-dependent energy resource. In general,
energy needs for a very wide variety of applications are also
time-dependent, but in a different fashion than the solar energy
supply. Consequently, the storage of energy or another product
of the solar process is necessary, if solar energy is to meet
these energy needs.

Energy (or product) storage must be considered in the light
of a solar process system, the major elements of which are:
the solar collector, storage units, conversion devices (such as
air conditioners or engines), loads, auxiliary or supplemental
energy supplies, and the control systems. The characteristics
and performance of each of these elements is related to that of
the others. The dependence of the collector performance on
temperature makes the whole system performance sensitive to temperature. For example, in a solar-thermal power system, a thermal energy storage system which is characterized by high drop in
temperature between its input and output will lead to high collector temperature and poor collector performance, low-heat engine inlet temperature and poor engine performance, or both.

The optimum capacity of an energy storage system depends
on the expected time-dependence of solar radiation availability,
the nature of loads to be expected on the process, the degree
of reliability needed for the process, the manner in which auxiliary energy is supplied, and an economic analysis that determines how much of the total (usually annual) loads should be
carried by solar and how much by the auxiliary energy source.
In this chapter we consider general questions of energy storage,
and in later chapters we will discuss particular applications
in house heating, water heating, and air conditioning.

Because of the difficulty of separating storage performance
from the collector and the load, illustrative problems are difficult to devise. The use of the concepts developed in this
chapter is demonstrated in subsequent chapters when we consider
complete systems.

9.1 PROCESS LOADS AND SOLAR COLLECTOR OUTPUTS

Consider a hypothetical solar process in which the time-dependence of the load L and gain from the collector Q_u are as shown in Figure 9.1.1a. During part of the time, available energy exceeds the load on the process, and at other times it is less. A storage system can be added to store the excess of Q_u over L when $Q_u > L$, and return it when $L > Q_u$.

Figure 9.1.1b shows, for this hypothetical process, the value of $\int (Q - L)d\tau$ as a function of time. The "zero time" is arbitrary, and its selection, in effect, only adds or subtracts a constant to all values of the integral. Over a given time period, the difference between the maximum and minimum (other than the initial value) of the integral represents the amount of storage capacity that would be required (assuming no losses) to permit the solar energy supply Q_u to meet all of the loads L.

In practice, it is usually not realistic to meet all of the loads on a process from solar energy over long periods of time, and an auxiliary energy source must be used. Where auxiliary energy is used, information such as that illustrated by Figure 9.1.1a shows the time dependence of auxiliary input.

It is also useful to show separately the integrated values of the major parameters Q_u, L, and A. Examples of these are shown in Figure 9.1.1c. The differences in the integrated values at any point in time are the same as the integrated difference shown in Figure 9.1.1b. A major use of information on long-term values of Q_u (and A) is to assess the cost of delivering energy or product from the solar energy process, and to estimate the fraction of total energy or product needs met from solar and auxiliary energy sources.

Information of this type can be generated over a short time (hours or days) or over a long time (months) by experiments or by modeling and simulation methods. The ordinates can represent energy or other products. The computation of the rates Q_u, L, and A can be carried out for the system by methods that will be outlined in following chapters, and are based on the component models outlined in this chapter and the previous chapters.

9.2 ENERGY STORAGE IN SOLAR PROCESS SYSTEMS

Energy storage may be in the form of sensible heat of a solid or liquid medium, as heat of fusion in chemical systems, or as chemical energy of products in a reversible chemical reaction. Mechanical energy can be converted to potential energy and stored in elevated fluids. Products of solar processes other

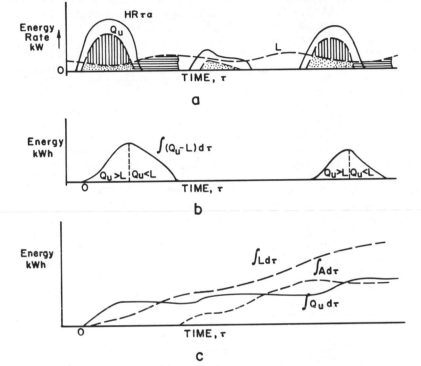

Figure 9.1.1 A hypothetical solar energy process with storage.
(a) Absorbed solar energy, $HR\tau\alpha$, collector useful gain, Q_u and
loads, L as a function of time for a 3-day period. Vertical
shaded areas show times of excess energy to be added to storage.
Horizontal shaded areas show energy withdrawn from storage to
meet loads. Dotted areas show energy supplied to load from collector during collector operation. (b) Energy added to or removed from storage, taking time $\tau = 0$ as a base. (c) Integrated
values of: useful gain from the collector, $\int Q_u d\tau$; load, $\int L d\tau$;
and auxiliary energy, $\int A d\tau$, for the same 3-day period. In this
example, for this period, solar energy collected is slightly
more than half the integrated load, and exceeded the auxiliary
energy supply.

217

than energy may be stored; for example, distilled water from a
solar still may be stored in tanks until needed.

The choice of media for energy storage depends on the nature of the process. For water heating, energy storage as sensible heat of stored water is logical. If air heating collectors are used, storage in sensible or latent heat effects in
particulate storage units are indicated, such as sensible heat
in a pebble bed heat exchanger. If photovoltaic or photochemical processes are used, storage is probably most logically in
the form of chemical energy.

A designer generally has alternatives in locating the energy storage component. As an example, consider a process in
which a heat engine converts solar energy into electrical energy; storage can be provided as thermal storage between the solar
collector and the engine, as mechanical storage between the engine and the generator, or as chemical storage in a battery between the generator and the end application. Solar cooling
with an absorption air conditioner provides another example.
Thermal energy can be stored from the collector to be used by
the air conditioner when needed or alternatively, the cooling
produced by the air conditioner can be stored in a low-temperature (below ambient) thermal storage unit. This and other
similar examples are illustrated in Figure 9.2.1.

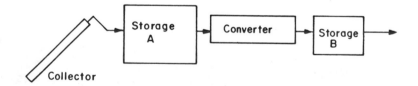

Figure 9.2.1 Schematic of alternative storage location, at A
or B.

These two alternatives are not equivalent in capacity,
costs, or effects on overall system design and performance.
The capacity required in a storage unit in position B is less
than that required in position A by (approximately) the efficiency of the intervening converter. Thus the capacity of B
must be only about 25% of the capacity of A if the conversion
process is operating at 25% efficiency. Thermal energy storage
at A has the advantage that the converter can be designed to
operate at more nearly constant rate, leading to better conversion efficiency and higher use factor on the converter; it can

lower converter capacity requirements by removing the need for
operation at peak capacities corresponding to direct solar input. The choice between energy storage at A or at B may have
very different effects on the operating temperature of the solar
collector, collector size, and ultimately on cost. These argu-
ments may be substantially modified, depending on how auxiliary
energy is used.
     The major characteristics of a thermal energy storage sys-
tem are: (a) its capacity, per unit volume, or weight; (b) the
temperature range over which it operates, that is, the temperature at which heat is added to and removed from the system;
(c) the means of addition or removal of heat and the temperature differences associated therewith; (d) temperature stratification in the storage unit; (e) the power requirements for
addition or removal of heat; (f) the containers, tanks, or other
structural elements associated with the storage system; (g) the
means of controlling thermal losses from the storage system;
and (h) its cost.
     Of particular significance in any storage system are those
factors affecting the operation of the solar collector. The
useful gain from a collector decreases as its average plate temperature increases. A relationship between the average collector temperature and the temperature at which heat is delivered
can be written as

$$T_{\text{collector}} - T_{\text{delivery}} = \Delta T_{\substack{\text{transport from}\\ \text{collector to storage}}} + \Delta T_{\text{into storage}}$$

$$+ \ \Delta T_{\text{storage loss}} + \Delta T_{\text{out of storage}}$$

$$+ \ \Delta T_{\substack{\text{transport from}\\ \text{storage to ap-}\\ \text{plication}}} + \Delta T_{\text{into application}}$$

Thus, the temperature of the collector, which determines its
useful gain, is higher than the temperature at which the heat
is finally used by the sum of a series of temperature-difference
driving forces. An objective of overall system design, and
particularly of storage unit design, is to minimize or eliminate
these temperature drops.

9.3 WATER STORAGE

Water is an inexpensive, readily available, and useful material

in which to store sensible heat. Energy is added to and removed from this type of storage unit by transport of the storage medium itself, thus eliminating any ΔT between transport
fluid and storage medium. If systems are well designed, pumping costs should be small and can be calculated by conventional
methods. Water storage units and collectors may be operated by
natural circulation in domestic solar-water heater applications
(see Chapter 11), or forced circulation may be used. Here we
are concerned primarily with forced circulation systems such as
are shown in Figure 9.3.1.

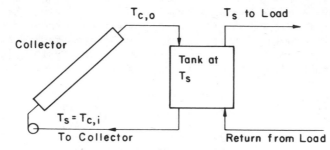

Figure 9.3.1 Water tank storage unit with water circulation
through collector to add energy, and through the load to remove
energy.

The energy storage capacity of a water (or other liquid)
storage unit operating over a finite temperature difference is
given by the usual heat capacity equation;

$$Q_s = (mC_p)_s (T_1 - T_2) \qquad (9.3.1)$$

where Q_s is the total heat capacity for a cycle operating between temperature limits T_1 and T_2, with m kilograms of water
in the unit. The temperature range over which such a unit can
operate is limited at the lower extreme (for most applications
by the requirements of the process, and at the upper limit by
the process or the vapor pressure of the liquid.

For a nonstratified tank, as shown in Figure 9.3.2, an energy balance on the tank yields

$$(mC_p)_s \frac{dT_s}{d\tau} = Q_u - L - (UA)_s (T_s - T_a) \qquad (9.3.2)$$

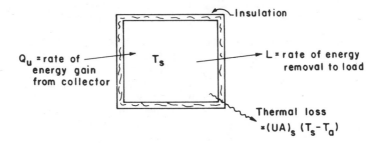

Figure 9.3.2 Unstratified storage of mass m, operating at time dependent temperature T_s in ambient T_a.

where Q_u and L are rates of addition or removal of energy from the collector and load.

Example 9.3.1 illustrates how the energy balances on a storage tank can be used to predict its temperature as a function of time. In this example, loads and input from the collector are given as functions of time, and the dependence of Q_u on storage-tank temperature is not shown. Examples in later chapters will illustrate this dependence.

> *Example 9.3.1* A fully mixed water tank storage containing 1500 kg of water has a loss coefficient-area product of 40 kJ/hr °C. The tank starts a particular 24-hr period at 45 °C and is in a room at a constant temperature of 20 °C. Energy Q_u is added to the tank from a solar collector, and energy is extracted from the tank to meet a load L. The first column of the table below indicates the time at the end of an hourly period; the second and third columns indicate values of Q_u and L for those hours. Calculate, using Euler integration, the temperature of the tank through the 24-hr period.
>
> *Solution.* The energy balance on the tank is represented by Eq. (9.3.2), which can be rewritten, for a finite increment in time, as
>
> $$T_{s,\text{new}} = T_{s,\text{old}} + \frac{\Delta\tau}{(mC)_s} [Q_u - L - (UA)_s(T_{s,\text{old}} - T_a)]$$
>
> Inserting the appropriate constants, with a time increment

of one hour,

$$T_{s,new} = T_{s,old} + \frac{1}{1500 \times 4.19} [Q_u - L - 40(T_{s,old} - 20)]$$

With this approximation, the temperature of the tank at the end of an hour is calculated from its temperature at the beginning of that hour, from the known inputs and outputs, and assuming that the loss term can be assumed constant throughout that hour. This has been done in Table 9.3.1, where column 4 is the temperature $T_{s,old}$ and column 5 is the new temperature for the end of that hour calculated from the equation.

TABLE 9.3.1 Work Table for Example 9.3.1.*

Hour	Q_u kJ $\times 10^{-3}$	L kJ $\times 10^{-3}$	$T_{s,old}$ °C	$T_{s,new}$ °C
1	0	12	45	42.9
2	0	12	42.9	40.8
3	0	11	40.8	38.9
4	0	11	38.9	37.0
5	0	13	37.0	34.8
6	0	14	34.8	32.5
7	0	18	32.5	29.6
8	0	21	29.6	26.2
9	21	20	26.2	26.3
10	41	20	26.3	29.6
11	60	18	29.6	36.2
12	75	16	36.2	45.5
13	77	14	45.5	55.4
14	68	14	55.4	63.8
15	48	13	63.8	69.1
16	25	18	69.1	69.9
17	2	22	69.9	66.4
18	0	24	66.4	63.2
19	0	18	63.2	59.7
20	0	20	59.7	56.3
21	0	15	56.3	53.7
22	0	11	53.7	51.7
23	0	10	51.7	49.9
24	0	9	49.9	48.3

*Note. $T_{s,new}$ for one hour becomes $T_{s,old}$ for the next hour.

Equation (9.3.2) can be rewritten to include the temperature and flow rates of collector and load streams. Neglecting any temperature drops of the fluid between tank and collector, Q_u can be written as

$$Q_u = (\dot{m}C_p)_c (T_{c,o} - T_s) \qquad (9.3.3)$$

where $(\dot{m}C_p)_c$ is the flow rate times specific heat of the fluid through the collector, and $T_{c,o}$ is the outlet fluid temperature; Q_u is calculated from collector Eq. (7.7.5). This expression for collector performance is particularly useful, as it is based on the collector inlet temperature, that is, on T_s which becomes the primary dependent variable.

The collector flow rate $\dot{m}$ can be thought of as the actual flow rate through the collector at any time; or, it can be viewed as the rate of pumping when the pump is operating. With this interpretation, Eq. (9.3.3) can be modified:

$$Q_u = F(\dot{m}C_p)_c (T_{c,o} - T_s) \qquad (9.3.4)$$

where F is a control function having a value of unity when the pump operates, and a value of zero at other times. The control function is thus a convenient analog of the "on" or "off" outputs of the controller on the pump, which turns the pump on when $T_{c,o} > T_s$ (i.e., when energy can be added to the storage unit). Using this concept, the flow rate at any time is $F\dot{m}$.

Similar relationships can usually be written for a load. The results are a set of equations, with Q_u determined by the collector performance equation and L by load requirements. These equations can be solved, usually numerically, to obtain T_s and various energy quantities as functions of time. Auxiliary energy input can also be included, if it is added into the tank or into the stream leaving the tank to the load. Solutions of these sets of equations are topics for the following chapters.

Water tanks may also operate with significant degrees of stratification, that is, with water not at uniform temperature over the vertical dimension of the tank. In this case, energy balances similar to Eq. (9.3.2) can be written for sections of the tank. This simulates the real situation on which the incoming water seeks its own density level, provided that it enters at a low velocity. For a two-section tank, as shown in Figure 9.3.3, an energy balance for the top section of the tank can be written

$$\frac{dT_{s,1}}{d\tau} = \frac{1}{(mC_p)_{s,1}} [F_1 (\dot{m}C_p)_c (T_{c,o} - T_{s,1}) + (\dot{m}C_p)_L (T_{s,2} - T_{s,1})$$
$$- (UA)_{s,1} (T_{s,1} - T_a)]$$

$$(9.3.5)$$

Here the first term in the brackets is the gain from the collector, multiplied by a control function, F_1, which is unity when
$T_{c,o}$ is higher than $T_{s,1}$ and 0 when $T_{s,1} > T_{c,o} > T_{s,2}$. The
second term represents the portion of the load supplied by the
top section of the tank. The return from the load for this tank
section model will always be to the bottom section. The total
load will be $(\dot{m}C_p)_L (T_{s,1} - T_{L,r})$, with an amount $(\dot{m}C_p)_L$
$(T_{s,1} - T_{s,2})$ supplied by the top tank and $(\dot{m}C_p)_L (T_{s,2} - T_{L,r})$
supplied by the bottom tank. The last term in the brackets is
the loss term from the upper section to the surroundings.

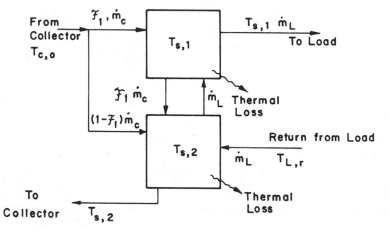

Figure 9.3.3 Partially stratified water storage tank, each
section considered to be of uniform temperature.*

An energy balance for the second section is

*This assumes some mixing between sections where both $\dot{m}_L$ and
$F_1 \dot{m}_c$ are finite. An alternative model would show one stream
between sections as the difference between $F_1 \dot{m}_c$ and $\dot{m}_L$, and
would represent maximum stratification.

$$\frac{dT_{s,2}}{d\tau} = \frac{1}{(mC_p)_{s,2}} \left[F_1 (\dot{m}C_p)_c (T_{s,1} - T_{s,2}) \right.$$

$$+ (1 - F_1)(\dot{m}C_p)_c (T_{c,o} - T_{s,2})$$

$$\left. + (\dot{m}C_p)_L (T_{L,r} - T_{s,2}) - (UA)_{s,2}(T_{s,2} - T_a) \right]$$

$$(9.3.6)$$

Note that terms will appear or will not appear in Eqs. (9.3.5) and (9.3.6) depending on temperatures and flow rates in the system. The collector flow rate $\dot{m}_c$ will be zero when the collector does not operate and the load flow rate $\dot{m}_L$ will be zero when the load does not operate; when both operate, some mixing occurs.

The equations for a two-segment tank can be generalized into an n-section tank if we define two control functions; one for the collector side and one for the load side. For the collector, we can define F_i^c such that

$$F_i^c = \begin{cases} 1 \text{ if } T_{i-1} > T_{c,o} > T_i \\ \\ 0 \text{ otherwise} \end{cases} \qquad (9.3.7)$$

and for the load, we can define F_i^L such that

$$F_i^L = \begin{cases} 1 \text{ if } T_i > T_{L,r} > T_{i+1} \\ \\ 0 \text{ otherwise} \end{cases} \qquad (9.3.8)$$

With these definitions of F_i^c and F_i^L, the energy balance for section i of an n-section tank is

$$(mC_p)_i \frac{dT_i}{d\tau} = (\dot{m}C_p)_c \left[F_i^c (T_{c,o} - T_i) + (T_{i-1} - T_i) \sum_{j=1}^{i-1} F_j^c \right]$$

$$+ (\dot{m}C_p)_L \left[F_i^L (T_{L,r} - T_i) + (T_{i+1} - T_i) \sum_{j=i+1}^{n} F_j^L \right]$$

$$+ U_i A_i (T_a - T_i)$$

$$(9.3.9)$$

Sheridan et al. (1967) have solved simple cases of stratified
storage tanks using an analog computer, while Gutierrez et al.
(1973) have solved a number of cases involving water heating on
a hybrid computer. The differences in predicted system performance using an unstratified model as compared to a stratified
model may be significant, depending on the nature of the application. These differences are most pronounced when a two- or
three-segment tank is substituted for a one-section tank. When
more than three sections are used in water heating simulations,
it was found by Gutierrez et al. that the predicted system performance did not change significantly from the three-section
simulations. Further study is needed to compare the results of
these analyses with experiments.

There may be circumstances under which the fluid in the
collector (or load) is not the same as in the storage tank. In
cold climates, for example, a fluid of low freezing point may
be circulated through the collector. A fully mixed storage
tank with heat exchangers in the tank is shown in Figure 9.3.4.

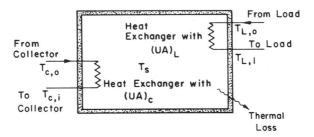

Figure 9.3.4 Storage tank of uniform temperature T_s with heat
added via heat exchanger c, and removed via exchanger L.

The basic energy balance for the tank is still Eq. (9.3.2),
but the rate of energy addition must be calculated from

$$Q_u = \left(\dot{m}C_p\right)_c \left(T_{c,o} - T_{c,i}\right) \qquad (9.3.10)$$

where, from basic "Effectiveness-Number of Transfer Units" (ε-
NTU) heat exchanger relationships, the temperature difference
is found from

$$\frac{T_{c,o} - T_{c,i}}{T_{c,o} - T_s} = 1 - e^{-(UA)_c/(\dot{m}C_p)_c} \qquad (9.3.11)$$

The removal of energy to the load is given by

$$Q_L = (\dot{m}C_p)_L (T_{L,i} - T_{L,o})$$ (9.3.12)

where the temperature difference across the load can be found
from

$$\frac{T_{L,i} - T_{L,o}}{T_s - T_{L,o}} = 1 - e^{-(UA)_L/(\dot{m}C_p)_L}$$ (9.3.13)

9.4 PACKED BED EXCHANGER STORAGE

A packed bed (pebble bed or rock pile) storage unit uses the
heat capacity of a bed of loosely packed particulate material
through which a fluid, usually air, is circulated to add or remove heat from the bed. A variety of solids may be used, rock
being the most widely used material.

Well-designed packed beds using rocks have several characteristics that are desirable for solar energy applications:
the heat transfer coefficient between the air and solid is high;
the cost of storage material is low; the conductivity of the
bed is low when air flow is not present.

A schematic of a packed bed storage unit is shown in Figure 9.4.1. Essential features include a container, a porous
structure to support the bed, and air distributors for flow in
both directions to minimize air channeling. In operation, flow
is maintained through the bed in one direction during addition
of heat (usually downward), and in the opposite direction during removal of heat. Note that heat cannot be added to this
storage unit and removed from it at the same time; this is in
contrast to water storage systems where simultaneous addition
to and removal from storage is possible. Insulation requirements at the outer surface of the packed bed are minimal, for
short-term storage, as the thermal conductivity of the bed in
the radial direction is low.

Pebble bed exchangers have the characteristic of very good
heat transfer between air and the solids of the bed; this tends
to minimize temperature differences from air to solids on heating the bed, and solids to air on cooling the bed. Many studies
are available on the heating and cooling of packed beds in chemical systems in which the packing material is of uniform size
and shape, but few consider materials of interest in solar systems. One study by Löf and Hawley (1948) investigated a range
of variables of interest in energy storage and arrived at the

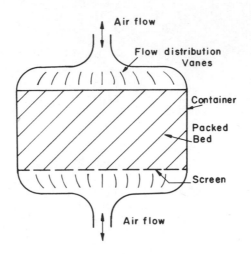

Figure 9.4.1   Schematic of a packed bed storage unit.

correlation

$$h_v = 650 \left[\frac{G}{D}\right]^{0.7} \tag{9.4.1}$$

where h_v is a volumetric heat transfer coefficient in W/m³ °C, G is the superficial mass velocity in kg/S m², and D is the equivalent spherical diameter of the particles in meters given by

$$D = \left[\frac{6}{\pi} \times \frac{\text{net volume of particles}}{\text{number of particles}}\right]^{\frac{1}{3}} \tag{9.4.2}$$

A well-designed bed has particle size small enough to minimize temperature gradients in the particles. The Biot criterion for spheres of radius R and thermal conductivity k can be applied; if the Biot number* hR/k is less than 0.1, thermal resistance within the particles (and thus internal temperatures gradients) can be considered negligible.[†] In experimental and

*h in the Biot number is the usual heat transfer coefficient per unit area; h is related to h_v of Eq. (9.4.1) by $(A/v)h = h_v$ where (A/v) is the surface area of the bed material per unit volume of bed material.

[†]Handley and Heggs (1969) suggest that internal temperature gradients can be accounted for by reducing h_v.

practical packed beds, rock sizes from about 1 to 5 cm have been used. Particle sizes should be uniform enough to obtain large void fractions and thus minimize pressure drop. Pressure drops are calculated by standard methods, such as outlined by Bird et al. (1960), and others. A useful summary of pressure drop and heat transfer considerations has been prepared by Close (1965).

Analytical or numerical approaches can be used to study the performance of packed bed exchangers, but obtaining an analytical solution for an arbitrary time-dependent inlet air temperature to the bed is very difficult. Since the output of a solar collector is a function of time through a day's operation, numerical methods are the most practical.

Consider a packed bed of total length L that is divided into N equal sections of length Δx. If we assume the radial temperature gradients are negligible, then within section i, the bed material can be approximated as having a single uniform temperature, $T_{b,i}$. This implies that the bed particle Biot number is less than 0.1. For small to moderate values of the volumetric heat transfer coefficient, the fluid temperature will not be the same as the bed temperature, and it will be necessary to write two energy balances; one for the bed and one for the fluid. For the ith bed section as shown in Figure 9.4.2 an energy balance for bed heating (air flow down) yields

$$(\rho C A \Delta x)_b \, \frac{dT_{b,i}}{d\tau} = h_v A \Delta x (T_{f,i-1} - T_{b,i}) - Q_{loss,i} \tag{9.4.3}$$

where A is the bed cross-sectional area, ρ is an apparent bed density, and C is the bed material specific heat. The heat transfer between the fluid and the bed was approximated by using the fluid temperature *into* the ith air section.

The loss to the surroundings on a unit area basis can be represented by an overall loss coefficient times the temperature difference between the bed surrounding air temperature. The surrounding temperature for storage loss is usually not the outside air temperature since for systems such as space heating, the storage unit will be located inside the dwelling. Storage losses then become uncontrolled gains to the dwelling.

If the heat capacity of the fluid is negligible compared to the bed heat capacity (the fluid is usually air), the fluid temperature out of the ith section, $T_{f,i}$, is given by*

$$(\dot{m} C_p)_c (T_{f,i-1} - T_{f,i}) = h_v A \Delta x (T_{f,i-1} - T_{b,i}) \tag{9.4.4}$$

*For numerical stability, the choice of Δx must be small enough so that $h_v A \Delta x / (\dot{m} C_p)_c < 1$.

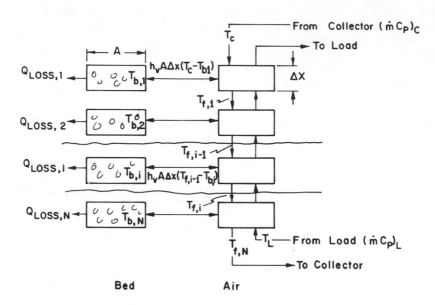

Figure 9.4.2 Numerical approximation to packed bed.

or

$$T_{f,i} = T_{f,i-1} - \frac{h_v A \Delta x}{(\dot{m} C_p)_c} (T_{f,i-1} - T_{b,i}) \qquad (9.4.5)$$

Equations (9.4.3) and (9.4.5) represent two sets of N equations (one set of algebraic equations and one set of differential equations) for the N unknown bed and N unknown fluid temperatures.

When energy is being extracted from the bed, a similar set of equations can be derived. It should be pointed out that the fluid flow rate from the collector may not be the same as from the load, and h_v for bed heating may not be the same for bed cooling.

When the volumetric heat transfer coefficient is large (but the Biot number is still less than 0.1), the temperature difference between the bed and fluid becomes small. The fluid temperature leaving a fluid section, $T_{f,i}$, will then be equal to the local bed temperature, $T_{b,i}$, and Eqs. (9.4.3) and (9.4.4) can be combined to yield

$$(\rho CA\Delta x)_b \frac{dT_{b,i}}{d\tau} = (\dot{m}C_p)(T_{b,i-1} - T_{b,i}) - Q_{loss,i} \quad (9.4.6)$$

Equation (9.4.6) represents N equations for the N unknown bed (and fluid) temperatures.

Pebble bed storage has been used in the Denver solar house, as described by Löf et al. (1964). Figure 9.4.3 indicates the temperature of the collectors on this house, for a particular day, while Figure 9.4.4 shows the corresponding temperature gradients in one of the two storage tubes as a function of the time of day. (This house heating system will be discussed in more detail in Chapter 12.)

9.5 PHASE-CHANGE ENERGY STORAGE

Materials that undergo a change of phase in a suitable temperature range may be useful for energy storage if several criteria can be satisfied. The phase change must be accompanied by a high latent heat effect and it must be reversible over a very large number of cycles without serious degradation.

The phase change must occur with limited supercooling. Means must be available to contain the material and transfer heat into it and out of it. And finally, the cost of the material and its containers must be reasonable. If these criteria can be met, phase-change energy storage systems can have high capacities (relative to energy storage in specific-heat type systems) when operated over small temperature ranges, with substantial reductions in volume and weight.

The earliest of the phase-change storage units to be studied experimentally for house heating applications [Telkes (1955)] was $Na_2SO_4 \cdot 10\ H_2O$, which decomposes at about 32 °C to give a solution plus Na_2SO_4 with a heat of fusion of 243 kJ/kg. The reaction is

$$Na_2SO_4 \cdot 10\ H_2O + Energy \rightleftharpoons Na_2SO_4 + 10\ H_2O$$

Energy storage is accomplished by the reaction proceeding from left to right on addition of heat. The total energy added depends on the temperature range over which the material is heated since it will include sensible heat to heat the salt to the transition temperature, heat of fusion to cause the phase change, and sensible heat to heat the Na_2SO_4 and solution to the final temperature. Energy extraction from storage is the reverse procedure, with the reaction proceeding from right to left and the thermal effects reversed.

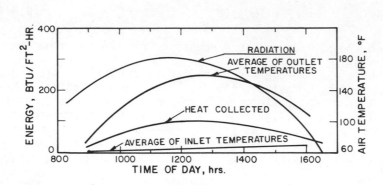

Figure 9.4.3 Collector performance for a November day for the Denver solar house, showing time dependence of collector output, i.e., storage unit input. [From Gillette (1959).]

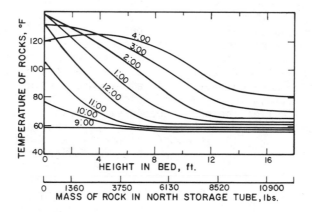

Figure 9.4.4 Temperature profiles in the packed bed storage unit for the same day as Figure 9.4.3. [From Gillette (1959).]

    Practical difficulties have been encountered with this system. It has been found that performance degrades on repeated cycling, with the thermal capacity of the system reduced. As shown in Figure 9.5.1a, $Na_2SO_4 \cdot 10 H_2O$ has an incongruent melting point, and as its temperature increases beyond the melting point it separates into a liquid (solution) phase and solid

Na_2SO_4. Since density of the salt is higher than the density
of the solution, a phase separation occurs. Attempts have been
made to use gels or other agents to avoid phase separation.

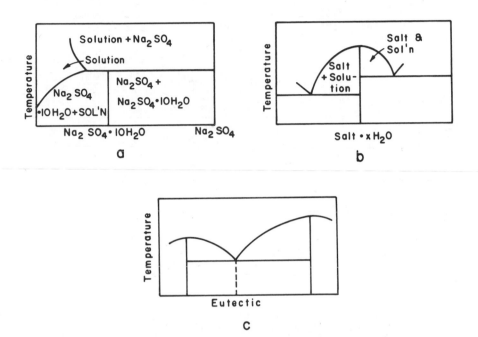

Figure 9.5.1 Phase diagrams of three types of systems of potential interest in thermal energy storage systems. (a) Part of
the Na_2SO_4-H_2O phase diagram, with the compound $Na_2SO_4 \cdot 10 H_2O$
with an incongruent melting point. (b) Part of a phase diagram
for a system (such as $FeCl_3$-H_2O) with a congruent melting point.
(c) Part of a phase diagram for a system which shows a eutectic
mixture. Adapted from Belton and Ajami (1973).

A range of other possibilities exist. A single material
with a suitable melting point avoids this major problem. The
transition from ice to water is an excellent historical example
that has been used for "energy" storage for centuries. Paraffin
waxes have recently been considered as possible energy storage
media. Compounds with congruent melting points such as
$Fe(NO_3)_2 \cdot 6 H_2O$, melt in the same manner as pure compounds.
Eutectics are also being considered; Kauffman and Gruntfest

(1973) list a range of possible eutectics, examples of which are indicated in Table 9.5.1. Phase diagrams for systems with congruent melting points are shown in Figure 9.5.1*b* and eutectic systems in Figure 9.5.1*c*.

TABLE 9.5.1 Examples of Eutectic Mixtures of Possible Interest in Thermal Energy Storage. [from Kauffman and Gruntfest (1973)].

Composition, wt %*	Melting Point, °C	Latent Heat	
		kcal/kg	kJ/kg
$CaCl_2-MgCl_2-H_2O$ 41-10-49*	25	41.7	175
$Mg(NO_3)_2 \cdot 6H_2O-Al(NO_3)_3 \cdot 9H_2O$ 53-47*	61	35.4	148
Acetamide-stearic acid 17-83*	65	52	218
$Urea-NH_4NO_3$ 45.3-54.7*	46	41	172

A further consideration with phase-change storage materials lies in the possibility of supercooling on energy recovery. If the material supercools, the latent heat of fusion may not be recovered or it may be recovered at a temperature significantly below the melting point. This question has been approached from three standpoints, by selection of materials that do not have a strong tendency to supercool, by addition of nucleating agents, and by ultrasonic means of nucleation. These considerations are reviewed by Belton and Ajami (1973), who note, for example, that the viscosity at the melting point of a material is a major factor in determining the glass forming ability of a melt and, thus, its tendency to supercool.

Heat transfer to and from a phase-change material must be given careful consideration. The material must be contained such a way that heat can be transferred to and from the material with a minimum temperature drop. This has been done experimentally by placing the material in small containers (cylindrical cans, tubes, or trays) with the containers placed in bins or ducts. The heat transfer fluid (usually air) is circulated

over the containers as in a packed bed. The heat transfer problem external to the containers is similar to a packed bed. In-
ternally, two additional phenomena must be considered; first,
the latent heat must be considered, which effectively adds a
high specific heat over a very small temperature range; and,
second, the thermal resistance to heat transfer within the material is variable with the degree of solidification and whether
heating or cooling of the material is occurring. [Heat transfer
in situations of this type have been studied, for example, by
Hodgins and Hoffman (1955) and Murray and Landis (1959).] As
heat is extracted from a phase-change material, crystallization
will occur at the walls and then progressively inward into the
material; at the end of the crystallization, heat must be transferred across layers of solid to the container walls. As a
solidified material is heated, melting occurs first at the walls
and then inward toward the center of the container.

Other practical factors must be taken into account, such as
corrosion, side reactions, vapor pressures, toxicity, and cost.

9.6 CAPACITIES OF STORAGE MEDIA

Ultimate comparisons of storage media and methods cannot be
accomplished without consideration of the total solar process
including the characteristics of the associated solar collectors,
loads on the process, the probable weather cycles, costs, and
many other factors. Some tabular data on the properties of
potentially useful storage media are shown in Table 9.6.1. The
data are not sufficient in themselves to judge the relative
merits of the media, but they give an indication of capacities,
weights, and volumes required for storage of a given quantity
of energy over any temperature range. These data are available
from standard handbooks.

The data in the table are for energy storage media that
might be used in applications for heating and cooling of buildings. Identical considerations apply to energy storage at higher temperatures. Temperature ranges might vary from 90 to
150 °C (for storage for an air conditioning process) up to 800
to 1000 °C (for a heat engine application for a power system).
McAllan (1974) surveys possible high-temperature storage media.
A major point of difference may be the increased magnitude of
thermal losses from the storage unit when it operates at temperatures far above ambient, and the corresponding need for improved thermal insulation.

TABLE 9.6.1 Heat Capacity and Heat of Fusion Data for Energy
Storage Media in Temperature Range of Operation of Flat-Plate
Collectors.

Material	C_p, kJ/kg °C	ΔH, Fusion, kJ/kg	T, °C	Density kg/m³
Water (liquid)	4.19	--	--	1000
(ice)	2.2	334	0°	--
Rock (typical)	0.88	--	--	2500-3500
Iron	0.50	--	--	7860
$Na_2SO_4 \cdot 10H_2O(s)$ (solution)	--	215	31°	1460

9.7 *STORAGE SYSTEM CAPACITY*

Three major factors determine the optimum capacity of storage
systems for buildings. First is the cost of the storage unit,
which includes the costs of containers, the storage medium itself, the space in which it is located, and the cost of operating the storage unit (costs of moving the heat transfer medium).
The second is the effect that storage capacity has on the operation of the balance of the system, and particularly on the collector; a smaller storage unit operating at a higher mean temperature can result in reduced collector output relative to a
similar system with a larger storage unit. These first two factors will be discussed further in later chapters on specific applications. The general observation can be made that "short-
time" storage, to meet loads for periods of a few days or less,
has been indicated as the most economical in studies of building applications.

Third factor to be considered is thermal losses from the
storage unit. Heat will be transferred outward through the
walls of any storage unit at a rate depending on the temperature difference between the storage media and the surroundings.
The total of such energy transfer is also a function of the time
of energy storage. If energy storage is to be considered for
long periods, thermal losses may become of critical importance.
This question has been considered by Speyer (1959) who, like
others, concluded that for house heating, long-term storage appears economically impractical.

It should be noted that in some operations, notably house heating, the storage unit may be located within the space to which heat is being added. In this situation, "losses" from the storage unit to the surroundings are not in fact losses, but are uncontrolled energy transfer from storage to the space to be heated. The situation is more critical in house cooling systems if storage losses add to the cooling load.

9.8 ALTERNATIVE STORAGE METHODS

The preceding sections have dealt largely with thermal energy storage. We can also consider other forms of energy storage or for specific applications; the following sections note several of these alternatives.

It is possible to convert mechanical to potential energy and recover the potential energy on demand to provide storage for mechanical systems. Such a system, for example, can pump water into an elevated reservior during periods when solar radiation is available, and recover the energy by running the water through a turbine when energy is needed. This method adds two inefficiencies to the overall system: the pump and the turbine. The efficiency of the storage system will be the product of the two efficiencies. Potential energy storage is practiced on very small and very large scales, for clocks on the one hand and for storing output of power plants on the other.

Processes can be conceived in which photochemical decompositions are brought about by solar radiation. An example of this is the photochemical decomposition of nitrosyl chloride:

$$2 \text{ NOCl} + \text{photons} \rightarrow 2 \text{ NO} + \text{Cl}_2$$

The reverse reaction can be carried out to recover the energy of the photons entering the reaction. In this case, storage in the system would be of the product chemicals, and the storage unit would consist of containers for each of the products. [See Marcus and Wohlers (1960,1961,1964).]

Processes that produce electrical energy may have storage provided as chemical energy in electrical storage batteries or their equivalent. Several types of battery systems can be considered for these applications, including secondary or primary batteries or fuel cell storage systems. Present possibilities include lead-acid, nickel-iron and nickel-cadmium batteries. The efficiencies of these systems range from 60 to 80% (ratio of watt-hour output to watt-hour input), depending on the battery, for low discharge rates and moderate charge rates. It is also possible to electrolyze water with solar-generated electrical energy, store oxygen and hydrogen, and recombine in a fuel cell to regain electrical energy [see Bacon (1964)]. These storage systems are characterized by relatively high cost per kilowatt-hour of storage capacity, and can now be considered for low-capacity special applications such as auxiliary power supply for space vehicles, isolated telephone repeater power supplies, instrument power supplies, and so on.

9.9 *SUMMARY OF STORAGE CONSIDERATIONS*

The desirable characteristics that an energy storage unit should possess (in degree depending on the application) can be summarized as follows:

1. The unit should be capable of receiving energy at the maximum rate without excessive driving forces,(i.e., temperature differences).
2. The unit should be capable of discharging energy at the maximum anticipated rate without excessive driving force.
3. The unit should have small losses (i.e., a low self-discharge characteristic).
4. The unit should be capable of a large number of charge-discharge cycles without serious dimunition of capacity.
5. The unit must be inexpensive.

The question of energy storage cannot be completely separated from that of the use of auxiliary energy supply. Analysis of thermal performance of systems costs of solar equipment and costs of auxiliary (conventional) energy can be used to determine the optimum size of the collector and the storage unit for a particular application and, thus, the relative amounts of solar and energy that should be provided.

REFERENCES

Bacon, F. T., Proceedings of the UN Conference on New Sources of Energy, *1*, 174 (1964). "Energy Storage Based on Electrolyzers and Hydrogen-Oxygen Fuel Cells."

Belton, G. and Ajami, F., Report NSF/RANN/SE/GI27979 TR/73/4 of the University of Pennsylvania National Center for Energy Management and Power to NSF (1973). "Thermochemistry of Salt Hydrates."

Bird, R. B., Stewart, W. C., and Lightfoot, E. N., Transport Phenomena, New York, Wiley, 1960.

Close, D. J., Mech. and Chem. Engr. Trans. Inst. Engrs., Australia, *MC1*, 11 (1965). "Rock Pile Thermal Storage for Comfort Air Conditioning."

Gillette, R. B., M. S. Thesis in Mechanical Engineering, Madison, University of Wisconsin (1959). "Analysis of the Performance of a Solar Heated House."

Gutierrez, G., Hincapie, F., Duffie, J. A., and Beckman, W. A., Solar Energy, *15*, 287 (1974). "Simulation of Forced Circulation Water Heaters; Effects of Auxiliary Energy Supply, Load Type, and Storage Capacity."

Handley, D. and Heggs, P. S., Int. J. Heat Mass Transfer, *12*, 549 (1969). "The Effect of Thermal Conductivity of the Packing Material on the Transient Heat Transfer in a Fixed Bed."

Hodgins, J. W. and Hoffman, T. W., Can. J. Technol., *33*, 293 (1955). "The Storage and Transfer of Low Potential Heat."

Kauffman, K. and Gruntfest, I., Report NCEMP-20 of the University of Pennsylvania National Center for Energy Management and Power, to NSF (1973). "Congruently Melting Materials for Thermal Energy Storage."

Löf, G. O. G. and Hawley, R. W., Ind. Eng. Chem., *40*, 1061 (1948). "Unsteady State Heat Transfer Between Air and Loose Solids."

Löf, G. O. G., El-Wakil, M. M., and Chiou, J. P., Proceedings of the UN Conference on New Sources of Energy, *5*, 185 (1964). "Design and Performance of Domestic Heating System Employing Solar Heated Air—The Colorado House."

Marcus, R. J. and Wohlers, H. C., Proceedings of the UN Conference on New Sources of Energy, *1*, (1964). "Chemical Conversion and Storage of Concentrated Solar Energy." See also: Solar Energy, *5*, 121 (1961); *5*, 44 (1961); *4*, No. 2, 1 (1960).

McAllan, J. V., Paper presented at Australia/U. S. Solar Energy Workshop at Sydney and Melbourne (1974). "Storage of High Grade Energy."

Murray, W. D. and Landis, F., Trans. ASME, J. Heat Transfer, *81C*, 107 (1959). "Numerical and Machine Solutions of Transient Heat Conduction Problems Involving Melting or Freezing."

Sheridan, N. R., Bullock, K. M., and Duffie, J. A., Solar Energy, *11*, 69 (1967). "Study of Solar Processes by Analog Computer."

Speyer, E., Solar Energy, *3*, No. 4, 24 (1959). "Optimum Storage of Heat with a Solar House."

Telkes, M., <u>Solar Energy Research</u>, Madison, University of Wisconsin Press, 1955. "Solar Heat Storage."

10. SOLAR PROCESS MODELS

In the preceding chapters, we have presented mathematical models for solar collectors and energy storage units. With these two major component formulations and with consistent mathematical models of other components in the system, plus additional information on the nature of the system loads and the driving forces, it is possible to represent the thermal performance of a solar process. The solution of the resulting set of equations provides useful information on the expected dynamic behavior of the system.

The technique is to write an energy balance equation for each component of the system. In general, the equations will be coupled so that it is necessary to find the simultaneous solution. For example, the collector performance is expressed in terms of the temperature of the fluid entering the collector. This in turn, for many systems, is the same as (or is closely related to) the temperature in the exit portion of the storage unit. The outlet temperature from the collector becomes the inlet temperature to the storage unit. In these equations time is the independent variable and the solution is in the form of temperatures as functions of time. In addition, it is possible to integrate energy quantities over time and thus assess the energy distributions for the system. This approach can be used to estimate, for a particular process application, the amount of energy delivered from the solar collector to meet a load and the quantity of auxiliary energy required. The simulation also indicates whether the temperature variations for a particular system design are reasonable, for example, whether a collector temperature would rise above the boiling point of the liquid being heated.

The ability to predict dynamic behavior and make energy balances over time is useful in several respects. The analysis provide a means of estimating the effects of design-variable changes on system performance; these design variables might for example, include selectivity of the absorbing surface, number of covers on the collector, collector area, and so on. In this context the system analysis provides a means of understanding how these systems function and can be a guide for experiment. Together with cost data, the system analysis can be used for

240

optimization to find the least-cost system, and it is thus a
design tool.

Several studies of system modeling, simulation, and optimization techniques of solar processes have been reported.
Gupta and Garg (1968) developed a model for thermal performance
of a natural circulation solar water heater with no load, represented solar radiation and ambient temperature by Fourier
series, and were able to predict a day's performance in a manner that agreed substantially with experiments. Gupta (1971)
used a response factor method that can be amenable to hand calculation for a day's process operation. Buchberg and Roulet
(1968) developed a thermal model of a house heating system,
simulated its operation with a year's hourly meteorological
data, and applied a pattern search optimization procedure in
finding optimum designs. Other process simulations have been
done by Löf and Tybout (1973) and Butz (1973) and will be noted
in Chapter 12.

In this chapter we provide a brief review of component
models and then demonstrate how these can be assembled into
system models. Examples of system analysis are presented which
provide an introduction to the following chapters on water heating, space heating, and air conditioning.

10.1 COMPONENT MODELS

Chapters 7 and 8 present a development of collector models and
Chapter 9 discusses storage unit models. For flat-plate collectors Eq. (7.7.5) is appropriate, while for focusing, Eq. (8.6.8)
can be used. Repeating the flat-plate equation

$$Q_u = A_c F_R [S - U_L (T_{c,i} - T_a)] \qquad (10.1.1)$$

where the useful gain can be related to the collector outlet
temperature by

$$Q_u = (\dot{m} C_p)_c (T_{c,o} - T_{c,i}) \qquad (10.1.2)$$

Implicit in this equation is a controller. Operation of a
forced circulation collector will not be carried out when
$Q_u < 0$, (or when $Q_u < Q_{min}$, where Q_{min} is a minimum level of
energy gain to justify pumping the fluid through the system).
In real systems, this is accomplished by comparing the temperature of the fluid leaving the collector (i.e., in the top header) with the temperature of the fluid in the exit portion of
the storage tank and running the pump only when the difference

in temperatures is positive and energy can be collected. This control function can be included explicitly by

$$Q_u = F(\dot{m}C_p)_c(T_{f,o} - T_{f,i}) \qquad (10.1.3)$$

where F is the control function having the value of unity when energy can be collected and zero at other times.

The basic equation for the performance of a fully mixed sensible heat storage unit is (9.3.2)

$$Q_u - L - (UA)_s(T_s - T_a) = (mC_p)_s\frac{dT_s}{d\tau} \qquad (10.1.4)$$

The equivalent equations for stratified water tank storage systems, pebble bed exchangers, or heat of fusion systems can be used in lieu of Eq. (10.1.4).

Equations (10.1.1) and (10.1.4), or variations on them, are the essential equations to be solved in most system analysis, but there may be additional components to be modeled such as heat exchangers, auxiliary energy supply systems, and so on. Additionally, the time dependences of S, the absorbed solar radiation per unit area of collector, and other essential weather data, must be known or estimated for the time period of interest.

The other term for which information must be available is L, the time rate of energy removal from the system. The determination of L may itself be a substantial problem. It can be a relatively simple matter of delivering a quantity of water at a minimum temperature for a service hot water requirement (with the hot water removed from the tank replaced by cold water from mains), or L can be energy transferred into the generator of an absorption air conditioner, with energy requirements fixed by characteristics of the air conditioner and the building. The general approach to determination of L is the same as that for other components; that is, develop the set of equations which relate energy rates and temperatures to time, which can be solved simultaneously with the component equations for collectors and storage units.

10.2 SYSTEM MODELS

System models are the collections of appropriate component models. The net effect of this collection is to produce either a set of all ordinary differential equations or a set of algebraic and ordinary differential equations, having time as the independent variable. These equations include meteorological data as forcing functions that operate on the collector, and that may also operate on the load, depending on the application. These equations can be manipulated and combined algebraically or they can be solved simultaneously without formal combinations. Each procedure has some advantages in solar process simulation. If the equations are all linear (and from a practical standpoint there are not too many to manipulate) the algebraic equations can be solved and substituted into the differential equations, which can then be solved by standard methods [Hamming (1962)]. If the algebraic equations are nonlinear, or if there are a large number of them coupled together so that they are difficult to solve, it is advantageous to leave them separated and solve the set of combined algebraic and differential equations.

As an example of a simple system that yields all differential equations, let us consider a solar water heater with an unstratified storage unit having a load that withdraws water at a fixed flow rate and returns water back to the tank from a constant temperature source. The collector equation can be combined with the storage tank equation to give

$$(mC_p)_s \frac{dT_s}{d\tau} = A_c F_R[S - U_L(T_s - T_a)] - (UA)_s(T_s - T_{room})$$
$$- (\dot{m}C_p)_L(T_s - T_{L,r})$$

$$(10.2.1)$$

Once the collector parameters, the storage size and loss coefficient, the magnitude of the load, and the meteorological data are specified, then the storage tank temperature can be calculated as a function of time. Also, gain from the collector, losses from storage, and energy to load can be determined for any desired period of time by integration of the appropriate rate quantities.

A number of methods are available to integrate equations like (10.2.1). In an example below, we integrate using simple Euler, which was the technique used in Example 9.3.1. Care must always be exercised when using most integration schemes to insure that the method is not unstable for the desired time step and that reasonably accurate solutions are being attained. When performing hand calculation, both stability and accuracy can be problems. However, most digital computer facilities have subroutines that will solve systems of differential equations and automatically take care of stability and accuracy problems.

Example 10.2.1 The performance of the collector of Example 7.10.1 was based upon a constant supply temperature of 60 °C to the collector. Assume the collector area is 2 m^2 and is connected to a water storage tank containing 100 kg of water initially at 60 °C. The storage tank loss coefficient-area product is 3 kJ/hr °C and the tank is located in a room at 25 °C. Assume water is withdrawn to meet a load at a rate of 5 kg/hr and is replenished from the mains at a temperature of 15 °C. Calculate the performance of this system for the period from 7:00 to 18:00 using the collector and meteorological data as in Example 7.10.1.

Solution. Using the Euler method we express the temperature derivative $dT_s/d\tau$ as $(T_{s,\text{new}} - T_{s,\text{old}})/\Delta\tau$ and obtain an expression for the change in storage tank temperature for the time period in terms of known quantities.

$$T_{s,\text{new}} - T_{s,\text{old}} = \frac{\Delta\tau}{(mC_p)_s} \{A_c F_R[S - U_L(T_{s,\text{old}} - T_a)]$$

$$- (UA)_s(T_{s,\text{old}} - T_{\text{room}}) - (\dot{m}C_p)_L(T_{s,\text{old}} - T_{L,r})\}$$

For convenience we will express this as

$$T_{s,\text{new}} - T_{s,\text{old}} = \frac{\Delta\tau}{(mC_p)_s} \{Q_u - Q_{\text{env}} - Q_{\text{load}}\}$$

For this problem a time step of one hour is sufficient to guarantee stability.

The stepwise solution is obtained by use of the proper parameters in the equation, which becomes (in terms of kilojoules),

$$T_{s,\text{new}} = T_{s,\text{old}} + \frac{1}{100 \times 4.19}$$

$$\{2 \times 0.824[S - 8 \times 3.6(T_{s,\text{old}} - T_a)]$$

$$- 3(T_{s,\text{old}} - 25) - 5 \times 4.19(T_{s,\text{old}} - 15)\}$$

The hourly steps in the solution are shown in Table 10.2.1, where the columns have the following meaning:

1. Column 1 is time.
2. Column 2 is the temperature of the storage tank at the start of the hour. It is found by adding column 9 to the previous tank temperature.
3. Column 3 is ambient air temperature.
4. Column 4 is absorbed solar radiation, in the hour period from Example 7.10.1.
5. Column 5 is collector gain.
6. Column 6 is hourly efficiency (HR is from Ex. 7.7.1).
7. Column 7 is storage loss.
8. Column 8 is load.
9. Column 9 is the sum of 5, 6, and 7.
10. Column 10 is the temperature change for the hour interval.

TABLE 10.2.1.

Time, Hrs	T_s,old, °C	T_a, °C	S, kJ/m²	Q_u kJ	η	Q_{env}, kJ	Q_L, kJ	ΣQ, kJ	ΔT, °C
6-7	60	0	0	--	--	105	943	-1048	-2.5
7-8	57.5	1	0	--	--	98	890	-988	-2.4
8-9	55.1	0	850	--	0	90	840	-930	-2.2
9-10	52.9	2	1380	--	0	84	794	-878	-2.1
10-11	50.8	4	2170	1360	0.26	77	750	+533	+1.3
11-12	52.1	10	3060	3040	0.46	81	777	+2181	+5.2
12-13	57.3	10	3130	2910	0.38	97	886	+1927	+4.6
13-14	61.9	8	2760	1990	0.30	111	983	+897	+2.1
14 15	64.0	8	2190	950	0.18	117	1027	-198	-0.5
15-16	63.5	6	1420	--	0	116	1016	-1131	-2.7
16-17	60.8	4	0	--	--	107	960	-1067	-2.5
17-18	58.3	1	0	--	--	100	907	-1007	-2.4
Totals			16,960	10,250		1183	10,773	-1709	

The final temperature is 55.9 °C. The change in internal energy is 100 × 4.19 × (55.9 - 60) = -1719 kJ, which is

nearly equal to the total ΣQ (as it should be if the calculation is correct).

The daily collector efficiency is $10250/(2 \times 3.6 \times 5883)$
$= 0.24$, which is considerably below the maximum hourly
efficiency of 0.42.

The load in this example is specified as a fixed flow rate,
regardless of water temperature. If a pattern of energy
delivery versus time had been specified, the temperature
history could have been quite different.

The preceding example was simple enough that hand calculation was possible to simulate a few hours of real time. Most
problems in solar simulation are not so easy nor are we usually
interested in only a few hours of simulated data. In general,
it is necessary to use a digital computer to obtain solutions.

Since most solar simulations involve basically the same
component models with only the size or other design variables
changing for different simulations, it is convenient to develop
a general simulation method. A program called TRNSYS has been
developed at the University of Wisconsin by Klein et al. (1974)
to simulate transient solar systems with a minimum of program-
ming effort. In this program general subroutines are available
that represent all the physical pieces of equipment that might
be necessary in an actual experiment. The only requirement is
instructing the program as to how the components are connected
together (by pipes and wires) and what are the values of the
basic design parameters of each component. The program does
the necessary simultaneous solution of algebraic and differential equations.

The integration algorithm selected for TRNSYS is the
Modified-Euler method. It is essentially a first-order predictor-corrector algorithm using Euler's method for the predicting step and the trapezoid rule for the correcting step.
The advantage of a predictor-corrector integration algorithm
for solving simultaneous algebraic and differential equations
is that the iterative calculations occurring during a single
time step are performed at a constant value of time. (This is
not the case for the Runge-Kutta algorithms.) As a result, the
solutions to the algebraic equations of the system converge, by
successive substitution, as the iteration required to solve the
differential equations progresses. The calculation scheme can
be described in the following manner.

At time τ, the values of the dependent variables, T^p, are
predicted using their values and the values of their derivatives,
$(dT/d\tau)_0$, from the previous time step:

$$T^p = T_0 + (\Delta\tau)\left(\frac{dT}{d\tau}\right)_0 \qquad (10.2.2)$$

where T^p is the predicted value of all of the dependent variables at time τ (note that this predict step is exactly the method used to integrate Example 10.2.1);

T_0 is the value of the dependent variables at time $(\tau - \Delta\tau)$;

$(\Delta\tau)$ is the time step interval at which solutions to the equations of the system model will be obtained; and

$(dT/d\tau)_0$ is the value of the derivative of the dependent variables at time $(\tau - \Delta\tau)$.

The predicted values of the dependent variables, T^p, are then used to determine corrected values, T^c, by evaluating their derivatives, $(dT/d\tau)$, as a function of τ, T^p, and the solutions to the algebraic equations of the model.

$$\frac{dT}{d\tau} = f(\tau, \; T^p, \; \text{algebraic solutions}) \qquad (10.2.3)$$

The corrected values of the dependent variables, T^c, are obtained by applying the trapezoidal rule:

$$T^c = T_0 + \frac{\Delta\tau}{2}\left[\left(\frac{dT}{d\tau}\right)_0 + \frac{dT}{d\tau}\right] \qquad (10.2.4)$$

If

$$\frac{2(T^c - T^p)}{(T^c + T^p)} > \varepsilon \qquad (10.2.5)$$

where ε is an error tolerance, then T^p is set equal to T^c and Eqs. (10.2.3) and (10.2.4) are repeated. When the error tolerance is satisfied, the solution for that time step is complete and the whole process is repeated for the next time step.

As an illustration of the use of the general program TRNSYS for simulating a solar energy system, and the nature of the results that can be obtained by simulations, consider the following example, shown schematically in Figure 10.2.1.

Example 10.2.2 A hot water load of 3000 kg of water per day at a minimum temperature of 60 °C is evenly distributed between the hours of 0700 and 2100. This load is to be met in substantial part by a solar collector assembly of total

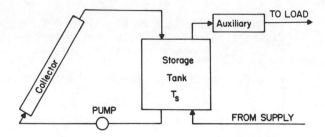

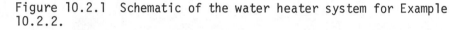

Figure 10.2.1 Schematic of the water heater system for Example 10.2.2.

effective area 65.0 m². The two-cover collector has the following characteristics:

Tilt = s = 40° to south

U_L = 14.4 kJ/hr °C m², or 4.0 W/m² °C

$(\tau\alpha)_e$ = 0.77

F' = 0.95

Flow rate through collector = $\dot{m}_c$ = 3250 kg/hr

The tank has the following characteristics:

V = 3.9 m³

Height/diameter ratio = 3

Loss coefficient U_L = 1.44 kJ/hr °C m²

Ambient temperature at tank = 21 °C

The supply water to the tank = 15 °C

The auxiliary heater is controlled such that if the temperature of the water from the tank is less than 60 °C, it will heat the water from the storage tank temperature to 60 °C. If T_s exceeds 60 °C, hotter water is delivered to meet the load, which is specified as a fixed volume of hot water.

The system is to be operated in Boulder, Colorado, latitude 40°N, for one week in January. Hourly values of solar radiation and ambient temperature are as shown in Figure 10.2.2 which are the same data as given in Table 3.3.3. Assuming that the initial tank temperature at the beginning

of the week is 60 °C, compute the percentage of the load
that is carried by solar energy.

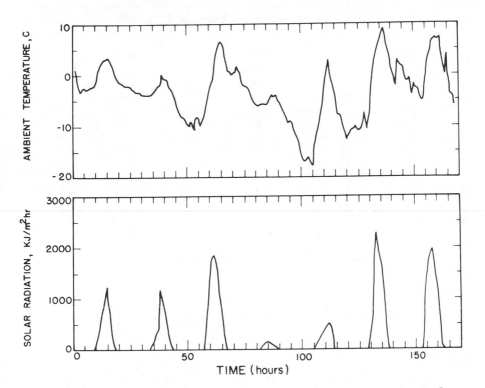

Figure 10.2.2 Air temperature and radiation on a horizontal
surface for Example 10.2.2.

The solution to this problem was obtained using TRNSYS and
a summary of the results are given in Table 10.2.2.

Two solutions are presented; one for an unstratified storage tank and one for a three-section storage tank. Since
the load requirement is a fixed amount of water of at least
60 °C, the two total loads are slightly different. Actually, the minimum total load is 3.960×10^6 kJ and both systems slightly exceed this value. The percentage of the
load carried by solar energy for the two cases are

$$\text{Case 1:} \quad \frac{2.74}{4.04} = 68\%$$

$$\text{Case 2:} \quad \frac{3.03}{4.10} = 74\%$$

TABLE 10.2.2 Summary of Results from Example 10.2.2.*

Case 1: Unstratified Storage Tank

End of Day	Total Incident Solar	Total Useful Gain	Total Tank Loss	Change of Energy of Tank	Load Supplied from Tank	Load Supplied from Auxiliary	Total Load
1	0.71	0.21	0.02	-0.27	0.46	0.10	0.56
2	1.34	0.43	0.03	-0.38	0.78	0.35	1.13
3	2.84	1.17	0.04	-0.11	1.24	0.47	1.71
4	2.95	1.17	0.05	-0.45	1.57	0.71	2.28
5	3.34	1.28	0.05	-0.53	1.76	1.00	2.76
6	4.87	2.09	0.07	-0.13	2.15	1.26	3.41
7	6.43	2.81	0.08	-0.01	2.74	1.30	4.04

Case 2: Three-Section Storage Tank

End of Day	Total Incident Solar	Total Useful Gain	Total Tank Loss	Change of Energy of Tank	Load Supplied from Tank	Load Supplied from Auxiliary	Total Load
1	0.71	0.24	0.02	-0.30	0.52	0.05	0.57
2	1.34	0.50	0.03	-0.42	0.89	0.25	1.14
3	2.84	1.28	0.04	-0.15	1.39	0.34	1.73
4	2.95	1.28	0.05	-0.58	1.81	0.50	2.31
5	3.34	1.42	0.05	-0.60	1.97	0.90	2.87
6	4.87	2.27	0.06	-0.18	2.39	1.06	3.45
7	6.43	3.05	0.08	-0.06	3.03	1.07	4.10

*All entries are integrated energy quantities $\times 10^{-6}$ (kJ).

These results, and results obtained with systems with larger collector and larger storage capacity for the same load and meteorological data are noted in section 11.9.

REFERENCES

Buchberg, H. and Roulet, J. R., Solar Energy, *12*, 31 (1968). "Simulation and Optimization of Solar Collection and Storage for House Heating."

Butz, L. W., M.S. Thesis in Mechanical Engineering, Madison, University of Wisconsin (1973). "Use of Solar Energy for Residential Heating and Cooling."

Gupta, C. L. and Garg, H. P., Solar Energy, *12*, 163 (1968). "System Design in Solar Water Heaters with Natural Circulation."

Gupta, C. L., Solar Energy, *13*, 301 (1971). "On Generalizing the Dynamic Performance of Solar Energy Systems."

Hamming, R. W. Numerical Methods for Scientists and Engineers, New York, McGraw-Hill, 1962.

Löf, G. O. G. and Tybout, R. A., Solar Energy, *14*, 253 (1973). "Cost of House Heating with Solar Energy."

Klein, S. A., Cooper, P. I., Beckman, W. A., and Duffie, J. A., Madison, University of Wisconsin, Engineering Experiment Station Report #38 (1974). "TRNSYS, A Transient Simulation Program."

11. SOLAR WATER HEATING

This is the first of four chapters on thermal energy for buildings. The second is on heating, and the third on addition of
absorption cooling to solar heating systems. The last is on
heating/cooling systems combining solar collectors and heat
pumps, systems that utilize solar energy for heating and nocturnal radiation for cooling, and on other systems that do not
directly fit the format of Chapters 11 to 13. The first parts
of these chapters include descriptions of systems, components,
and important design considerations. Published descriptions
and data on several existing systems are reviewed. Then modeling and simulation methods are used to show prediction of performance and design of systems.
  Considerations important in designing water heating systems
are also basic to solar heating and cooling systems. These four
chapters should be studied as a group, with each succeeding
process being on a larger scale or operating at higher temperatures above ambient, and being thus of more critical design.
  In this chapter, we have drawn from publications of the
Mechanical Engineering Division of the Commonwealth Scientific
and Industrial Research Organization of Australia. Their Circular No. 2 entitled "Solar Water Heaters" (1964) is a useful
general reference, particularly for systems using natural circulation.

11.1 *WATER HEATER SYSTEMS*

The basic elements in the most common solar water heater are
the flat-plate collector and storage tank. These are connected
to supply a load, auxiliary energy is usually provided, and
means for circulation of water and control of the system must
be included. An example of a natural circulation system is
shown schematically in Figure 11.1.1.
  In this system, the tank is located above the collector,
and water will circulate by natural convection whenever solar
energy in the collector adds energy to the water in the collector leg and thus establishes a density gradient. Auxiliary is
shown here as added energy into the top of the tank to maintain

hot water in the tank top at some minimum temperature level necessary to meet the loads.

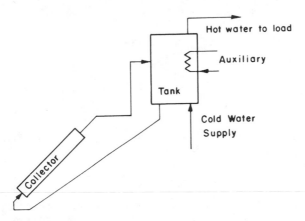

Figure 11.1.1 Schematic of a natural circulation solar water heater, with auxiliary energy added to the storage tank.

Figure 11.1.2 shows schematically an example of a forced circulation system. Here there is no requirement for location of the tank above the collector. A pump is required, which is usually controlled by a differential controller turning on the pump when the temperature at the top header is several degrees higher than the temperature of the water in the bottom of the tank. A check valve is needed to prevent reverse circulation and resultant nighttime thermal losses from the collector. In this example, auxiliary energy is shown as provided to the water leaving the tank and going to the load.

Solar water heaters have become widely used and are the basis of small manufacturing operations in Australia, Israel, and Japan. Many of the heaters manufactured in Japan have been of a unique design and will be noted at the end of the chapter; most of the Australian and Israeli household heaters utilize natural circulation; some larger heaters depend on forced circulation. Figure 11.1.3 shows domestic and institutional hot water systems in Australia.

11.2 COLLECTORS AND STORAGE TANKS

The flat-plate collectors most commonly used are shown

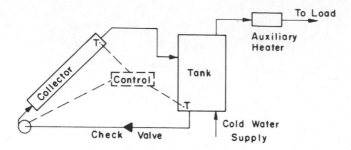

Figure 11.1.2 Schematic of a forced circulation solar water heater, with auxiliary on line to load.

schematically in Figure 11.2.1. The absorbers typically use parallel 1.2- to 1.5-cm diameter tubes, 12 to 15 cm apart, soldered or brazed into headers of about 2.5 cm in diameter at the top and bottom. The tubes are soldered or otherwise thermally bonded to the plates. The most common absorber plate material in Australia is copper; the Israeli Miromit heater uses galvanized iron. The absorber plates are mounted in a metal or asbestos cement box, with 5 to 10 cm of insulation behind the plate and one (and occasionally two) glass cover over the plates, leaving about 2.5-cm air gaps. The dimensions of a single collector are typically 1.2 m × 0.6 m or 1.2 m × 1.2 m; more than one collector may be used in an installation.

The performance of the collectors is outlined in Chapter 7. The configurations described above are of the same geometry on which the analysis is based. Other geometries have also been used, such as a single serpentine tube rather than parallel tubes, and plates formed of one flat and one corrugated sheet fastened together to form water passages. These are shown in Figure 11.2.2.

Some Australian heaters utilize selective surfaces prepared by a process developed by CSIRO [see Close (1962a)]. The Israeli heaters manufactured by Miromit have a Tabor selective surface applied to them [Sobotka (1964)].

Storage tanks must be well insulated, and common practice is to use about 20 cm of mineral wool insulation on sides, top, and bottom. The piping from collector to tank must also be well insulated, and arranged to minimize pressure drop. Piping of 2.5 cm or larger is typical for household units, with runs as short as is practical.

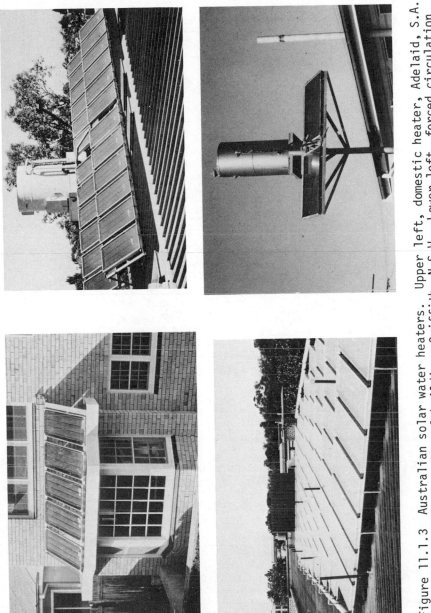

Figure 11.1.3 Australian solar water heaters. Upper left, domestic heater, Adelaid, S.A.
Upper right, heater on hostel building, Griffith, N.S.W. Lower left, forced circulation
water heaters, hostel, Darwin, N.T. Lower right, domestic heater, Mt. Isa, Qld. (Photos
courtesy of CSIRO.)

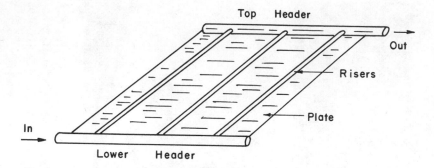

Figure 11.2.1 Schematic of a "conventional" solar water heater collector plate.

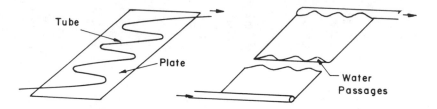

Figure 11.2.2 Alternative water heater collector plates.

It is advantageous to maintain stratification in the storage tanks, and thus the location and design of tank connections is important. The schematics in Figures 11.1.1 and 11.1.2 show approximate locations of connections in typical use. Close (1962b) shows data on tank temperatures at various levels in an experimental natural circulation water heating system operated for a day with no hot water removal from the tank. These data are shown in Figure 11.2.3; the degree of stratification is evident.

11.3 *LOADS AND SIZING OF SYSTEMS*

As with any other solar energy application, the optimum size of a solar water heater to meet a particular service hot water need (i.e., load) depends on a combination of factors, including the investments required in the solar system, the costs of alternative (auxiliary) energy, collector orientation, climate, and temperature of the cold water supply.

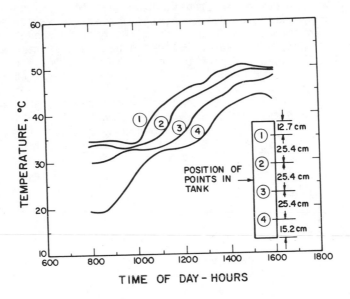

Figure 11.2.3 Temperature distribution with vertical position
in a storage tank with a natural circulation water heater.
[From Close (1962b).]

In the case of household hot water supplies, a body of experience has been accumulated on which designs are based, and
"rules of thumb" are used for system design. In Australia,
for example, systems are designed to produce water at 65 °C,
and at a daily average usage of 45 kilograms per person per day.
If an all-solar system is to be used, for example, in Darwin,
Australia (which is characterized by almost continuous good solar weather and expensive conventional fuels) a storage capacity
of 2.5 times the daily requirement is suggested. If an auxiliary energy source is to be used (e.g., in Melbourne, where radiation is more intermittent and conventional energy is less
expensive) the recommended tank size is about 1.5 times the
daily requirement. For a family of four, in Darwin, a collector area of about 4 m^2 is suggested. Collectors should be oriented approximately toward the equator at a slope of 0.9ϕ for
maximum annual collection; deviations of up to 22° east or west
of north (or south) should have small effect on annual performance.

The magnitude and time dependence of loads in institutional buildings may be easier to predict than that for residences, and additional effort at design of the systems is warranted by the larger investment in them. The illustrative problem of section 11.9 and (and Chapter 10) shows how such systems can be simulated. The load information needed in the simulation is the quantity of hot water needed (usually at temperatures above some minimum level) as a function of time, or the quantity of energy needed (measured relative to some reference temperature and at temperatures above a minimum) as a function of time. Figure 11.3.1 shows a schematic of time dependence of hot water needs of two buildings--a residence and a laboratory or office building.

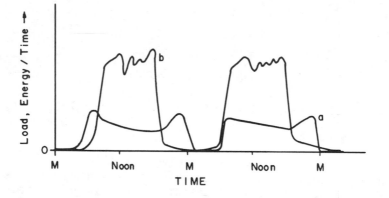

Figure 11.3.1 Schematic of time dependence of energy loads on hot water system: a) residential; b) laboratory or office.

Forced circulation systems are more commonly used in larger institutional installations. Collectors are larger assemblies of the individual modules described above. Davey (1970), in a survey of solar water heating in Australia, lists institutional systems, shown in Table 11.3.1.

11.4 AUXILIARY ENERGY

As with other solar energy applications, solar water heaters have a time-dependent output that is influenced by weather cycles. The degree of reliability desired of a process to meet a particular load can be provided by a combination of properly

TABLE 11.3.1 Areas and Storage Capacities of Forced Circulation Water Heaters in Australia [from Davey (1970)].

Location	Area, m^2	Capacity, liters
Plant research building--Canberra	31.2	1510
Church hostel--Alice Springs	26.0	1890
Church hostel--Alice Springs	18.6	1320
School--Mt. Isa	14.9	760
Hotel--Fiji	41.8	3670
Caravan Park--Exmouth Gulf two installations, each	33.6	2840
Caravan Park--Broken Hill	11.8	910
Airline staff quarters--Darwin nine installations, each	23.8	1510
Welfare hostel--Darwin	17.7	1140
Youth hostel--Darwin	35.7	1510
Hotel--New Guinea	14.9	1140
Hotel--New Guinea	41.8	3410
School--Adelaide	59.5	3410
Laboratory--Melbourne	17.8	950

sized collector and storage units, and an auxiliary energy source. In areas of very high solar energy availability (e.g., where there seldom are clouds of significant duration), it may be practical to provide total loads with the solar systems. In climates of lower radiation availability, as typical of temperate climates, auxiliary is needed to provide high reliability and avoid gross overdesign of the solar system and unproductive investment in it.

Auxiliary energy can be provided in three ways; (A) energy can be added to the tank as shown in Figure 11.1.1; (B) energy can be added to the water leaving the tank as in Figure 11.1.2; and, (C) in addition, energy can be added directly to the incoming feed water, bypassing the tank. The first two are the most widely used methods.

The relative merits of these three methods have been examined by Gutierrez et al. (1974). By simulating a month's operation of the three systems having fixed collector size and a stratified (3-node) storage tank under a range of conditions of ambient temperature, time of day, and magnitude of loads, the relative portion of the loads carried by solar and by auxiliary energy were estimated. It was shown that when the collectors operated at temperature ranges not greatly above ambient temperatures, the method of adding auxiliary was not critical, with method B showing minor advantages. However, when the minimum hot water temperature was raised 20 °C, method B showed significant advantages over method A, which in turn was better than method C. Examples of the fraction of the load carried by solar energy are shown in Table 11.4.1.

TABLE 11.4.1   Ratio of Auxiliary Energy to Total Load for 31-Day Period for a Solar Water Systems [from Gutierrez et al. (1974)].

	Auxiliary Method		
Type	A	B	C
N 750- 5- 90S	0.17	0.16	0.18
N1050- 5- 90S	0.32	0.31	0.35
N1500- 5- 90S	0.50	0.49	0.53
M 750- 5- 90S	0.34	0.33	0.35
M1050- 5- 90S	0.52	0.51	0.53
M1500- 5- 90S	0.66	0.66	0.67
N 750-10- 90S	0.12	0.11	0.12
N1050-10- 90S	0.26	0.24	0.26
N1500-10- 90S	0.46	0.44	0.46
M 750-10- 90S	0.20	0.18	0.21
M1050-10- 90S	0.35	0.33	0.35
M1500-10- 90S	0.53	0.51	0.54
N 750-20- 90S	0.09	0.08	0.10
N1050-20- 90S	0.20	0.18	0.21
N1500-20- 90S	0.39	0.38	0.43
M 750-20- 90S	0.14	0.12	0.15
M1050-20- 90S	0.27	0.24	0.27
M1500-20- 90S	0.46	0.43	0.47
N 750-10- 90W	0.16	0.15	0.16
N1050-10- 90W	0.33	0.31	0.34
N1500-10- 90W	0.52	0.51	0.53
M 750-10- 90W	0.26	0.23	0.27
M1050-10- 90W	0.41	0.40	0.42
M1500-10- 90W	0.60	0.58	0.60
N 750-10-110S	0.14	0.12	0.16

continued

TABLE 11.4.1 (Continued)

| Type | Auxiliary Method | | |
	A	B	C
N1050-10-110S	0.28	0.26	0.32
N1500-10-110S	0.49	0.45	0.50
M 750-10-110S	0.22	0.19	0.25
M1050-10-110S	0.37	0.33	0.40
M1500-10-110S	0.54	0.53	0.57
M 750-10-150S	0.30	0.24	0.39
M1050-10-150S	0.46	0.39	0.55
M1500-10-150S	0.60	0.54	0.68

Key: M: early morning loads; N: noon loads; magnitude of
load: 750, 1050 or 1500 Btu/day ft^2; storage capacity: 5, 10,
or 20 Btu/°F ft^2; minimum control temperature: 90, 110, or
150 °F; month: summer, S, or winter, W.

The major reasons for the changes in performance on changing the method of adding auxiliary concern the temperature at
which the collector operates. Adding auxiliary energy to the
top of the tank (method A) can result in a higher mean collector temperature, poorer collector performance, and higher requirements for auxiliary energy. Method C, which bypasses the
tank when its top section is not hot enough, results in failure
to use some collected solar energy. Method B with a modulated
auxiliary heater (or its equivalent in an on-off auxiliary heater) maximizes use of the solar collector output and minimizes
collector losses by operating at the lowest mean collector temperature of any of the methods.

11.5 FLOW DISTRIBUTION IN COLLECTORS

Standard design and performance calculations of collectors are
based on an implicit assumption of uniform flow distribution in
all of the risers in single or multiple collector units. If
flow is not uniform, some parts of the collectors with low flow
through risers may run at significantly higher temperature than
parts with higher flow rates. Thus the design of both headers
in individual collectors and manifold systems for multiple collectors is important in obtaining good collector performance.
This problem has been studied analytically and experimentally
by Dunkle and Davey (1970). It is of particular significance

in large forced-circulation systems; natural circulation systems
tend to be self-correcting and the problem is not as critical.
    Based on the assumptions that flow is turbulent in headers
and laminar in risers (assumptions logical for water heaters
that are built in Australia), the analysis by Dunkle and Davey
shows pressure drop along the headers for the common situation
of water entering the bottom header at one side of the collector and leaving the top header at the other side. To illustrate
the equations, they consider a bank of six typical solar collector absorbers, each 0.60 m × 1.20 m, with headers of 2.54 cm
diameter and risers of 1.3 cm diameter. The total header length
is 46 m, the viscosity of water is taken as 6.6×10^{-4} Ns/m^2,
the friction factor in the headers is 0.10, and the flow rate
is 218 kg/hr for the whole bank. Under these circumstances,
the calculated pressure distribution in the top and bottom
headers are as shown in Figure 11.5.1. The implications of
these pressure distributions are obvious; the pressure drops
from bottom to top are greater at the ends than the center portion, leading to high flows in the end risers and low flows in
the center risers.

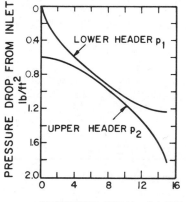

Figure 11.5.1 Calculated pressure distribution in headers of
an isothermal absorber bank [from Dunkle and Davey (1970)].

    This situation is found experimentally. Temperatures on
absorber plates are measures of how effectively energy is removed, and thus differences among temperatures measured at the
same relative location on individual collectors in banks is a

measure of the lack of uniformity of flow in risers. Figure
11.5.2 from Dunkle and Davey, shows measured temperatures for a
bank of twelve collectors connected in parallel. The data show
temperature differentials of 22°C from center to ends; these
are significant differences, as can be shown by reference to
collector energy balances. Connecting the units in a series
parallel or multiple parallel arrangements such as shown on
Figure 11.5.3 results in more uniform flow distribution and
temperatures.

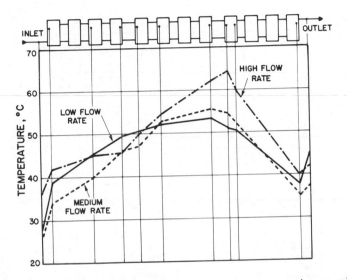

Figure 11.5.2 Experimental temperature measurements on plates
in a bank of parallel flow collectors, at three flow rates
[from Dunkle and Davey (1970)].

Dunkle and Davey recommend, as a result of this study, that
collectors be designed with headers large enough so that the
major pressure drop is in the risers. Banks up to 24 risers
can be satisfactory for either forced or natural circulation.
For forced circulation banks of over 24 risers, no more than 16
risers should be connected in parallel; for larger banks series
parallel or multiple parallel connections can be used.

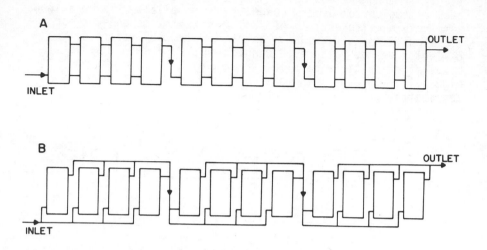

Figure 11.5.3 Alternative methods of connecting collectors in
banks: (a) Series-parallel; (b) Parallel-series [from Dunkle
and Davey (1970)].

11.6 PERFORMANCE OF NATURAL CIRCULATION SYSTEMS

Natural circulation in solar heaters such as that shown in Figure 11.1.1 occurs when the collector warms up enough to establish a density difference between the leg including the collector and the leg including the tank and the feed line from tank
to collector. The density difference is a function of temperature difference, and the flow rate is thus a function of the
useful gain of the collector which produces that temperature
difference. Under these circumstances, these systems are self-
adjusting, with increasing gain leading to increasing flow rates
through the collector. It has been observed by Löf and Close
(1967) and by Cooper (1973) that under wide ranges of conditions
the increase in temperature water through collectors in natural
circulation systems (particularly those of Australian design)
is about 10 °C.
     Close (1962b) worked out an analysis of circulation rates
in natural circulation systems and compared computed and experimental inlet and outlet temperatures. His results, some of
which are shown in Figure 11.6.1, confirm the suggestion that
temperature increases of about 10 °C are representative of these

systems if they are well designed and without serious flow re-
strictions. Gupta and Garg (1968) also show inlet and outlet
water temperatures for two collectors that suggest nearly constant temperature rise across the collectors.

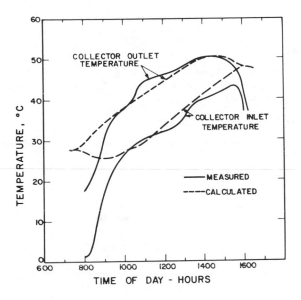

Figure 11.6.1 Collector inlet and outlet temperatures during
hours of sunshine for a natural circulation water heater [from
Close (1962b)].

    Thus there are two alternative methods of modeling the per-
formance of the collector in natural circulation systems. The
first is by an analysis of the temperature and density distributions and resulting flow rates based on pressure drop calculations (e.g., as outlined by Close).
    The second is to assume the simpler model of constant tem-
perature increase of water flowing through the collector of
10 °C (or other appropriate value) and calculate the flow rate
which will produce this ΔT_f at the estimated collector gain.
From Eq. (7.7.5),

$$Q_u = F_R A_c [S - U_L (T_{f,i} - T_a)] \qquad (11.6.1)$$

and

$$Q_u = \dot{m} C_p (T_{f,i} - T_{f,o}) = \dot{m} C_p \Delta T_f \qquad (11.6.2)$$

Equating these,

$$\dot{m} C_p = F_R A_c \frac{S - U_L (T_{f,i} - T_a)}{\Delta T_f} \qquad (11.6.3)$$

This equation can be solved for $\dot{m}$ if it is assumed that F' is independent of flow rate. Substituting Eq. (7.7.4) for F_R into Eq. (11.6.3) and rearranging gives

$$\dot{m} = - \frac{U_L F' A_c}{C_p \ln \{1 - (U_L F' \Delta T_f)/[S - U_L (T_{f,i} - T_a)]\}} \qquad (11.6.4)$$

Collector operation with ΔT_f of the order of 10 °C implies for practical systems that water circulates through the collector several times per day. Tabor (1969) suggested an alternative, that is, that resistance to flow be higher and ΔT_f higher, with flow rates such that the water in the tank makes one pass through the collector in a day. He calculated that the daily efficiency of a "one-pass" high ΔT_f system will be about the same as a system using several passes per day and lower ΔT_f, if better stratification is obtained in the tank. Many Israeli water heaters are designed to operate this way.

11.7 FREEZING TEMPERATURES

Ambient temperatures below freezing impose special problems on water heating systems, as collectors can be exposed to ambient conditions for periods of many hours without solar radiation. If this condition is expected, even as a rare event, three possible solutions may be considered:

1. The collectors may be designed to be drained during times when they are not operating or when freezing is a possibility.
2. Antifreeze solutions can be used in the collector, with heat exchanger to transfer the energy to the water.
3. The collectors may be designed to withstand occasional light freezes by use of devices such as butyl rubber top headers.

Most of the applications of water heaters have been made under circumstances where freezing is not a problem. In climates where freezing is a problem, the first two methods have been used experimentally. The use of antifreeze solution imposes an extra heat exchanger between collector and storage tank, by schemes such as those shown in Figure 11.7.1. This heat exchanger increases the temperature of operation of the collector by an amount corresponding to the temperature drop across the exchanger. As a rough rule of thumb, each degree Celsius temperature drop across the exchanger decreases the useful gain of the collector by 1 to 2%, and for the same desired solar energy output requires a correspondingly larger collector. Thus, effective heat exchanges should be used.

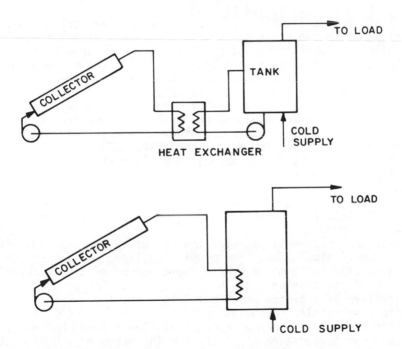

Figure 11.7.1 Methods for use of antifreeze solutions in water heater collectors.

11.8 TWO SPECIAL PURPOSE DESIGNS

There are two special solar water heaters worthy of note because their design or their performance is unique. In Japan, a large number of water heaters are built which combine the storage and collector in a single unit. For example, one design is comparable to a plastic air mattress of about 1.6 m² in area, is mounted horizontally, and has a capacity of about 200 liters. It is filled in the morning, and by the end of the day the water temperature has risen to the point where it is useful for evening domestic water needs [Tanishita (1964)]. More recent designs also combine collector and storage tank functions. Collectors with rows of glass and plastic tubes about 0.15 to 0.2 m in diameter, painted black, and enclosed in a box fitted with FRP or polycarbonate top cover, are now sold in significant numbers in Japan [Tanishita (1970)].

There are applications of solar water heaters for the warming of swimming pools. Separate storage tanks are not needed, and collectors operate relatively few degrees above ambient temperatures. Under these circumstances the collectors are usually operated by forced circulation without covers and with a minimum of insulation. They may be made of metal or plastic, and may, for example, be installed as a roof material for a patio or pool cabana. Swimming pool heaters have been described by Andrassy (1964), Boyd (1973), and others; deWinter (1973) has written a monograph on design and construction of pool heaters, and as of 1973 at least one manufacturer is marketing solar swimming pool heaters.

11.9 RESULTS OF SIMULATION OF A WATER HEATING SYSTEM

The previous chapter illustrated the use of system analyses and solar process simulation. The problem outlined was to compute the integrated performance over a short period of time (a week) of a forced circulation water heating system. The problem is illustrative of a system which might be used, for example, in a dormitory or similar installation.

The loads were taken as a constant flow rate of hot water at a minimum temperature of 60 °C from the hours of 0700 to 2100. The collector, storage tank, and controls were specified on the basis of reasonable practice and experience. Auxiliary energy is provided by method B of section 11.4.

The simulation of a week's operation is not adequate to obtain adequate design information but can show clearly the effects on winter performance of several interesting design variations, each from the original design. The variable used as an indicator of performance is the percentage of load supplied by solar energy. The tank (top) temperature is also of interest to see if the system increases to temperature levels above the boiling point.

The baseline design specified in the problem provided 68% of the load in the winter week; the maximum 6 p. m. tank temperature was 68 °C.

When stratification was assumed to be obtained in the tank, that is, when three nodes were assumed, the collector provided 74% of the load. The maximum tank temperature in the top was 74°, and at that time the bottom tank temperature was 55 °C.

When the collector area and storage capacity were doubled the collector provided 89% of the load, and the maximum temperature was 86 °C.

When the storage volume is doubled but the collector area remains the same, solar energy provided 71% of the load.

These results are intended to illustrate the effects of design changes on performance of a system and to show the kinds of information that can be obtained from simulations. Caution must be exercised in interpreting this example; the changes brought about by the large perturbations selected for the illustration cannot be expected to hold for other systems and for this system with smaller loads or under other meteorological conditions. The trends are shown, but the magnitude of the changes in amount of the load supplied by solar energy cannot be generalized from this example.

REFERENCES

Andrassy, S., Proceedings of the UN Conference on New Sources of Energy, *5*, 20 (1964). "Solar Water Heaters."

Boyd, T., Popular Science (May 1973).

Close, D. J., Report E.D.7, Engineering Section, Commonwealth Scientific and Industrial Research Organization, Melbourne, Australia (1962). "Flat-Plate Solar Absorbers: The Production and Testing of a Selective Surface for Copper Absorber Plates."

Close, D. J., Solar Energy, *6*, No. 1, 33 (1962). "The Performance of Solar Water Heaters with Natural Circulation."

Cooper, P. I., Personal communication (1973).

Davey, E. T., Paper presented at Melbourne International Solar Energy Society Conference (1970). "Solar Water Heating in Australia."

deWinter, F., "How to Design and Build a Solar-Energy Swimming Pool Heater," a publication of the Copper Development Association, Inc. (1973).

Dunkle, R. V. and Davey, E. T., Paper presented at Melbourne International Solar Energy Society Conference (1970). "Flow Distribution in Absorber Banks."

Gupta, C. L. and Garg, H. P., Solar Energy, *12*, 163 (1968). "System Design in Solar Water Heaters with Natural Circulation."

Gutierrez, G., Hincapie, F., Duffie, J. A., and Beckman, W. A., Solar Energy, *15*, 287 (1974). "Simulation of Forced Circulation Water Heaters; Effects of Auxiliary Energy Supply, Load Type and Storage Capacity."

Löf, G. O. G. and Close, D. J., <u>Low Temperature Engineering Application of Solar Energy</u>, New York, ASHRAE, 1967. "Solar Water Heaters."

Sobotka, R., Proceedings of the UN Conference on New Sources of Energy, *5*, 96 (1964). "Solar Water Heaters."

"Solar Water Heaters, Principles of Design, Construction and Installation," Div. of Mech. Engr. Circular No. 2, Commonwealth Scientific and Industrial Research Organization, Melbourne, Australia (1964).

Tabor, H., Bulletin, Cooperation Mediterraneene pour L'Eenergie Solaire (COMPLES), No. 17, 33 (1969). "A Note on the Thermosyphon Solar Hot Water Heater."

Tanishita, I., Proceedings of the UN Conference on New Sources of Energy, *5*, 102 (1964). "Recent Development of Solar Water Heaters in Japan."

Tanishita, I., Paper presented at Melbourne International Solar Energy Society Conference (1970). "Present Situation of Commercial Solar Water Heaters in Japan."

12. SOLAR HEATING

Heat for comfort in buildings can be provided from solar energy
by systems that are, in concept, enlarged versions of the water
heater systems described in Chapter 11. The two most common
heat transfer fluids are water and air, and systems based on
each of these are described in this chapter. The basic components are the collector, storage unit, load (i.e., the house or
building to be heated), and controls. In temperate climates,
an auxiliary energy source must be provided and the design problem is in part a problem of determination of the optimum combination of solar energy and auxiliary (i.e., conventional) energy.
     Buildings with large windows facing the equator (south in
the northern hemisphere, or north in the southern hemisphere)
and arranged to admit solar radiation into the building when the
sun is low in the winter sky, have been termed "solar houses."
The gains to be realized from properly oriented windows are significant, but in cold climates losses during periods of low radiation, nights, and cloudy weather, must be controlled so that
net gains can be realized. We shall not treat these matters,
but will discuss solar heating systems that provide controlled
heating in a manner comparable to fuel-fired systems.
     In this chapter we review only solar heating (in combina-
tion with solar heating of hot water). Chapter 13 deals with
the possibility of addition of cooling by the solar operation
of absorption (or mechanical) cooling cycles, using the same
collector that is used for winter heating and water heating.
Chapter 14 discusses other types of systems for solar heating
and cooling, such as combination with heat pumps or with use
nocturnal radiation to dissipate energy and provide cooling.
     A gift by Godfrey L. Cabot to Massachusetts Institute of
Technology for solar energy studies in 1938 initiated a solar
heating program that resulted in a series of four experimental
solar houses. Interest in solar heating in the United States
further developed during World War II, when the country faced
the possibility of prolonged hostilities and attendant oil
shortages. Although there was some earlier effort (e.g., a
patent by Morse in 1881), the basis for this chapter is work
since 1938.

271

In 1950 a symposium of Space Heating with Solar Energy was
held at MIT, and the papers at that symposium were published
under that title [Hamilton (1954)]. A more recent review by Löf
(1964), for the United Nations Conference on New Sources of
Energy in 1961, is a summary of 11 papers presented at that
conference, each describing buildings of various designs in
diversified locations and climates. There has been significant
new work since the Löf review, but so far it has not included
much operating data and experiments with full-scale systems.
Such data on new systems should soon become available. In this
chapter we do not suggest that we have included reference to all
of the important experiments and research, but have used and
referenced those that show how several solar heating systems
function, how they can be modeled and designed, and what the economics of the process may be.

12.1 SURVEY OF SOLAR HEATING EXPERIMENTS

A series of experiments at MIT involved the successive development of four solar heated structures, with the last of these at
Lexington, Massachusetts. This building was a carefully engineered and instrumented system based on solar water heaters
and water storage, as described by Engebretson and Ashar (1960)
and Engebretson (1964). The solar heating system was designed
to carry about two-thirds of the total winter heating loads in
the Boston area. The design and performance of this building
will be discussed in some detail in section 12.4.
    Telkes and Raymond (1949) described a solar house that was
constructed at Dover, Massachusetts, that utilized vertical
south-facing air heater collectors and energy storage in the
heat of fusion of sodium sulfate decahydrate. This system was
designed to carry the total heating load, having theoretical
capacity in the storage system to carry the design heating loads
for five days.
    Bliss (1955) constructed and operated a fully solar heated
house in the Arizona desert, using a matrix, through-flow air
heater and a rock pile energy storage unit. It was noted that
the system as built did not represent an economic optimum, and
a smaller system using some auxiliary energy would have resulted in lower cost.
    Löf designed an air heating system using overlapped glass-
plate collectors and rock pile exchangers for energy storage
and, using these concepts, built a residence near Denver, in
which he has lived since 1959. The performance of this system
during the first years of its operation was studied and reported
by Löf et al. (1963,1964), and is discussed in section 12.3.

Close, Dunkle, and Robeson (1968) described a heating system used for partial heating of a laboratory building in Australia that has been in operation for eight years. This system uses a 56 m^2 vee-groove air heater and a rock pile storage unit. Air flow through the collectors is modulated to obtain a fixed 55 °C air outlet temperature.

Water heating systems using large-capacity water storage tanks and rock beds surrounding the tanks for additional storage and heat transfer area are described by Thomason (1973). Böer (1973) and his associates have designed and built a solar heated structure at Newark, Delaware, that is designed to use photovoltaic cells as part of the energy-absorbing surface in the lower temperature portions of the flat-plate air heaters. Several solar heated houses have been constructed by the Solar Energy Laboratory of the Centre National de Recherche Scientifique at Odeillo, France, by Trombe and his associates. These houses utilize south walls of glass, behind which are black-painted concrete walls about 20 cm thick which serve both the energy-absorbing and storage functions. Air circulates around the concrete wall by natural convection to provide heat distribution to the rooms of the houses. More information on the Odeillo and Delaware houses is given in Chapter 14.

12.2 SOLAR HEATING SYSTEMS

The major components of solar heating systems are collector, storage, and auxiliary energy source. It is useful to consider these systems as having four basic modes of operation, depending on the conditions that exist in the system at a particular time:

Mode A: If solar energy is available and heat is not needed in the building, energy gain from the collector is added to storage.

Mode B; If solar energy is available and heat is needed in the building, energy gain from the collector is used to supply the building need.

Mode C: If solar energy is not available, heat is needed in the building, and the storage unit has stored energy in it, that stored energy is used to supply the building need.

Mode D; If solar energy is not available, heat is needed in the building, and the storage unit has been depleted, auxiliary energy is used to supply the building need.

Note that there is a fifth situation that will exist in

practical systems. That is, the storage unit can be fully
heated, there are no loads to be met, and the collector can gain
energy. Under these circumstances, there is no way to use or
store collected energy and this energy must be discarded. Additional operational modes may also be provided, for example, to
provide service hot water. It is also possible with some systems to combine modes, that is, to operate in more than one mode
at a time.

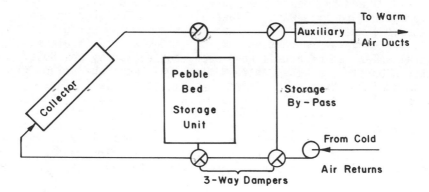

Figure 12.2.1 Schematic of basic hot air system.

        Figure 12.2.1 is a schematic of a basic air heating system
with a pebble bed storage unit and auxiliary furnace. In this
system the storage medium (pebbles) is held in the storage unit
while air is the fluid used to transport energy from collector
to storage and to the building. The four modes of operation
are achieved by appropriate damper positioning. With this system it is not possible to combine modes by both adding energy
to and removing energy from storage at the same time. The use
of auxiliary can be combined with energy supply to the building
from collector or storage if that supply is inadequate to meet
the loads.
        Figure 12.2.1 shows the blower on the upstream side of the
collector; this arrangement results in collectors operating at
pressures slightly above ambient. Blowers can also be placed
so that the pressure in the collectors is not above ambient
pressure, which may be advantageous in controlling leakage.
        Systems based on this concept have a number of advantages,
compared with those based on use of water as a heat transfer
medium. There is no problem with freezing in the collectors,
and problems of designing for overheating during periods of no
energy removal are minimized. Corrosion problems are also

minimized. The working fluid is air, and warm air heating systems are in common use. Conventional control equipment is readily available that can be applied to these systems. Disadvantages include relatively high fluid pumping costs (particularly
if the storage unit is not carefully designed), relatively large
volumes of storage, and the difficulty of adding conventional
absorption air conditioners to the systems.

Figure 12.2.2 is a schematic of a basic water heating sys-
tme, with water tank storage and auxiliary energy source. This
system allows independent control of the solar collector-storage
part of the system on the one hand, and storage-auxiliary-load
part of the system on the other, as solar heated water can be
added to storage at the same time that hot water is removed
from storage to meet building loads. In the system illustrated,
a bypass around the storage tank is provided to avoid heating
the storage tank with auxiliary energy.

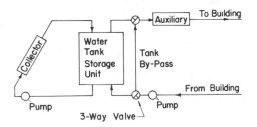

Figure 12.2.2 Schematic of basic hot water system.

Advantages of water heating include use of a common heat
transfer and storage medium (avoiding temperature drops to
transfer energy into and out of storage), smaller storage volume, relatively easy adaptation to supply of energy to absorption air conditioners, and relatively low-energy requirements
for pumping of the heat transfer fluid.

Use of water also involves problems. Freezing of collectors must be avoided, by means such as those outlined in Chapter 11 for water heaters. Solar water heating systems will
probably operate at lower water temperatures than conventional
water systems and thus require additional heat transfer area
or equivalent means to transfer heat into the building. Water
heaters may also operate at excessively high temperatures (particularly in spring and fall) and means must be provided to remove energy and avoid boiling and pressure build-up. Care must
also be exercised to avoid corrosion problems.

To illustrate air and water heating systems, the next sections describe the Löf house in Denver and the MIT House IV,
both of which have been carefully engineered, their performance
measured, and the results documented.

12.3 *THE DENVER SOLAR HOUSE*

The Denver house was constructed in 1958 to 1959, and has been
described in some detail in two publications by Löf, El-Wakil,
and Chiou (1963,1964). This discussion is a considerably ab-
breviated version of these papers. The house uses an air heater, pebble bed storage unit, and natural gas furnace for an
auxiliary source. The house has a floor area of about 2100 ft^2
(195 m^2) and an area of 1100 ft^2 (102 m^2) in the basement. The
building is of contemporary design with substantial areas of
windows and a flat roof. The design heat load was calculated
at 108,500 Btu/hr (114,500 kJ/hr) at ambient temperature of 0°F
(-18 °C) and wind speed of 8.8 mph (3.9 m/sec).
The heating system is shown schematically in Figure 12.3.1.
Collectors are in two banks, angled up from the flat roof at a
slope of 45°. Each bank has a nominal area of 300 ft^2 (27.9 m^2)
for a total collector area of 600 ft^2 (55.7 m^2). The effective
collector area, allowing for collector frames, and so on, is
530 ft^2 (49.2 m^2). Collectors are of the overlapped glass-
plate design noted in Chapter 7; air flows through two modules
in series, the first having one cover and the second having two.
The storage medium is 23,500 lb (10,640 kg) of rock of a
nominal mean size of 1 to 1.5 in. (3 cm) in diameter, and a
specific heat of 0.18 Btu/lb °F (0.75 kJ/kg °C). It is contained in two cylindrical tubes of 3 ft (0.91 m) in diameter and
18 ft (5.5 m) high. A duct leads down through one of the storage tubes as a means of access between collector banks on the
roof and equipment in the basement.
Some solar energy is provided for service hot water heating
by an air-water heat exchanger serving as a preheater. The balance of needed heat for hot water is then provided in a fuel-
fired conventional heater. Other major components are shown in
the diagram, such as the blower, furnace, dampers, cold and warm
air registers, and so on.
Solar radiation is sensed by a resistance thermometer
mounted on a black plate above and parallel to one of the collectors, which is designed to have a thermal response similar
to the black plate in the collector. Another resistance thermometer is located in the exit (in this case the top) of the
storage units, and air to the collector is turned on or off by

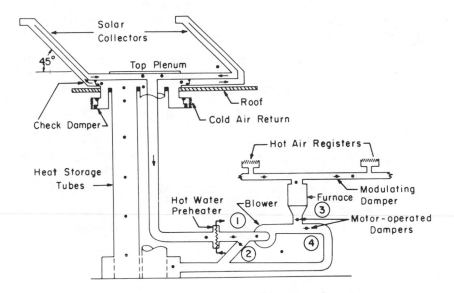

Figure 12.3.1 Schematic of the Denver House air heating system.
[from Löf et al. (1963)].

the detection of an appropriate temperature difference between
the two resistance elements.
      Room thermostats calling for heat in the house control
dampers that direct air from collector or storage to the house.
If, after a period of about 10 min of air flow from collector
or storage, the condition called for by the thermostat is not
met, the furnace turns on. When collectors are operating,
heated air goes to the building if needed; otherwise it goes to
storage.
      The four modes of operation of the previous section (and
combinations thereof) are used in this system:

      A. When the house does not need heat and solar energy is
high enough to justify collection, dampers 1 and 4 are open,
2 and 3 are closed. Air flow is from collector to water preheater to blower to storage unit and return to collector.
      B. When house needs can be met directly by solar energy,
dampers 1 and 3 are open, 2 and 4 closed. Air flow is from
collector to water preheater to blower to furnace to hot air
registers, through the building to cold air returns to collector.

Note that in this system some gas heat may be supplied by the furnace if solar heat is not sufficient to meet house demands.

   C. For house heating from storage, dampers 2 and 3 are open, 1 and 4 closed. Air flow is from rooms to cold air returns, down through storage to blower to furnace to hot air registers to room.

   D. Operation completely on auxiliary (gas) is the same as *C*, except that the furnace is on. Thus, in fact, some heat from storage can be supplemented by gas or all energy may come from gas.

   The results of measurements of performance of the system have been reported for the 1959 to 1960 heating season. They are based on measurements of air temperatures, air flow rates, storage temperatures, indication of modes of operation, and gas consumption for space heating and for hot water. With these data, energy quantities in various parts of the system were integrated for months and for the total heating season. Table 12.3.1, from Löf et al., shows the year's data for the Denver house. Figure 12.3.2 shows cumulative performance data for the heating season, and Figure 12.3.3 shows how the solar radiation incident on the collector was utilized or lost.

TABLE 12.3.1 Energy Balance of Denver Solar House Winter 1959 to 1960: all values in million Btu [from Löf et al. (1964)].

	Total
Total solar incidence on 600 ft^2 collector area	226.86
Total solar incidence on 600 ft^2 collector area when collection cycles operated	161.33
Useful collected heat	55.72
Net collector efficiency, percent	34.6
Solar heat absorbed by storage tubes	25.46
Solar heat absorbed by water preheater	3.95
Heat delivered by natural gas for house heating	141.83
Heat delivered by natural gas for water heating	20.43
Total heat load	217.98
Percent of useful collected heat absorbed by water preheater	7.1
Percent of total water heating load supplied by solar energy	16.2
Percent of house heat load supplied by solar energy (including water preheating but excluding water heating)	28.2
Percent of house heat load supplied by solar energy (including both water preheating and heating)	25.7

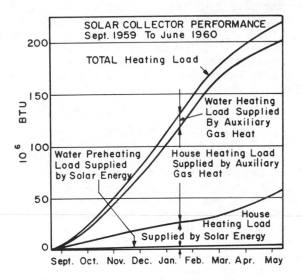

Figure 12.3.2 Cumulative total heating loads, auxiliary energy
used, and solar energy supplied for a heating season for the
Denver solar house [from Löf et al. (1964)].

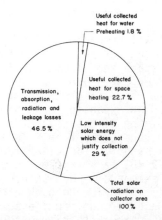

Figure 12.3.3 The distribution of the total solar energy incident on the Denver house collector, for the heating season, into
delivered energy and various losses [from Löf et al. (1964)].

These measurements are based entirely on thermal considerations; they do not reflect costs or in any way show what the
optimum collector size might be for that house. Based on analyses that show optimum fractions of solar energy in the range
of one-half to three-fourths, the annual one-fourth of the load
carried by solar energy in the Denver house is substantially
lower than the optimum for a full-scale design. This system
has been in routine operation since 1959 with only nominal maintenance, indicating its durability.

12.4 MIT HOUSE IV

MIT House IV has been described in detail by Engebretson and
Ashar (1960) and Engebretson (1964). It was built in the late
1950s, and data on its operation are available for the two heating seasons of 1959 to 1960 and 1960 to 1961. The house used
water in the collectors and storage tank, and an oil-fired auxiliary to supply space heating and hot water needs of the two-
story, 1450 ft^2 (135 m^2) building. The house was designed specifically to incorporate solar heating in the most effective
way. The collectors were part of the envelope of the building,
which had minimum surface-to-volume ratio. At the same time,
the house was designed to be a pleasant and functional structure. The collector was the dominant architectural feature (in
contrast to the design of the Denver house which used a relatively smaller collector that is partially hidden from view).
A schematic of the system is shown in Figure 12.4.1. The
collector had an area of 640 ft^2 (59.5 m^2) at a tilt of 60° to
the south, using two low-iron content glass covers, black-
painted aluminum absorbing surface, copper tubing in risers and
headers, and both air space and fiberglass insulation on the
back. For this collector, F_R = 0.86, α = 0.97, and U_L = 0.70
Btu/hr °F ft^2 (3.97 W/m^2 °C). The storage tank had a capacity
of 1500 gal (5670 kg). An expansion tank of 200 gal (757 kg)
capacity permitted draining of the collector system when it was
not in use. Heat was transferred to air in the house through a
water-to-air heater exchanger. The oil-fired auxiliary heater
included a 100-gal (379 kg) tank from which water was circulated
to the water-air heat exchanger. Hot water was supplied by
passing the city supply in series through a heating coil in the
main storage tank and another coil in the auxiliary tank; the
resulting hot water was mixed with city water to obtain the desired 140 °F water.
The collector pump was controlled by the difference in
temperature between a sensor element applied to the collector
plate, and the storage tank. Since this collector was drained

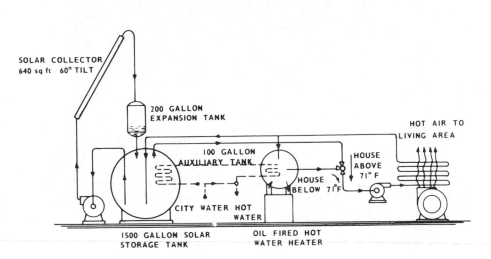

Figure 12.4.1 Schematic of the MIT House IV heating system
[from Engebretson (1964)].

when not in use, the control allowed a small amount of overheating of the collector before start-up, and also allowed some de-
lay in the form of thermal response lag in the collector sensor
to avoid premature shut-down when the first cool water from the
piping system entered the collector.

Space heating control was by means of a two-stage thermostat in the house. As the temperature in the room dropped, the
heat exchanger pump (and circulating fan) turned on to provide
heat from the storage tank. If the temperature continued to
drop, that is, if the solar energy stored could not meet the
load, the position of the motorized valve shifted to allow circulation of hot water from the auxiliary tank rather than the main
tank.

These modes of operation were somewhat different than those
for the air heating system, as the collector-storage part of the
system was controlled and operated independently of the storage-
load part. Thus, in the nomenclature of section 12.2, *Mode A*
operated whenever the energy incident on the collector was
enough to deliver useful energy to the fluid at the bottom storage tank temperature. *Mode B* did not exist in this system.
Mode C operated when the house needed heat and the storage tank
could provide it, the first stage of operation called for by the
room thermostat. *Mode D* operated on the second stage of the
thermostat, with heat delivered by the auxiliary heater.

   The house was instrumented with means for measuring appropriate temperatures and flow rates as a function of time, to
permit calculation of energy balances on all components and on
the system as a function of time. These energy balances were
integrated to give cumulative performance through the heating
seasons. The results of these measurements are shown in Figure
12.4.2 and 12.4.3 [from Engebretson (1964)], each of which shows
1959 to 1960 data (solid lines) and 1960 to 1961 data (dashed
lines). Figure 12.4.2 shows collector performance. Figure
12.4.3 shows cumulative performance of the house, indicating
cumulative building needs for space heat and for domestic hot
water heat, and how each of these was met by a combination of
solar and auxiliary energy. These figures show clearly that
the weather was significantly better and that the solar energy
delivered by this system to the house was substantially higher
in the second year than in the first.
   Table 12.4.1 summarizes some of the season total integrated
energy quantities for the building for the two years.

TABLE 12.4.1 Integrated Performance of MIT House IV over Two
Heating Seasons [from Engebretson (1964)].

	Btu × 10^6		kJ × 10^6	
	1959-60	1960-61	1959-60	1960-61
Space heating demand	68.7	67.0	72.5	70.7
Space heating from solar energy	31.8	38.1	33.6	40.2
Water heating demand	13.9	16.7	14.7	17.6
Water heating from solar energy	8.0	9.2	8.4	9.7
Total heating demand	82.6	83.7	87.1	88.3
Total heating from solar energy	39.7	47.3	41.9	49.9
Percent from solar energy	48.1	56.6	48.1	56.6

12.5 SOLAR HEATING ECONOMICS

The preceding discussion was concerned with the thermal performance of solar heating systems and did not deal with questions
of costs. The design of systems must be performed with due consideration of costs, and as an introduction to the following

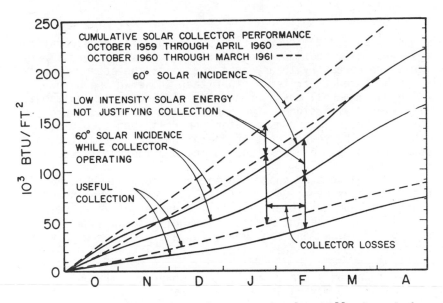

Figure 12.4.2 MIT House IV integrated solar collector performance [from Engebretson (1964)].

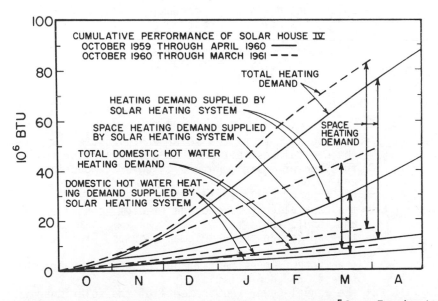

Figure 12.4.3 MIT House IV thermal performance [from Engebretson (1964)].

sections on system design we discuss some general considerations of costs of solar heating. Note that estimates of thermal performance are necessary to any cost analysis; the economic analysis may be based on criteria other than those we use here.

The major annual costs of a solar heating system, without auxiliary energy, include: the annual cost of owning the collector, storage unit and associated controls, pumps, piping, and the like; the yearly cost of operating the system; the cost of power for pumps, blowers, and so on; and the yearly cost of maintenance.

The annual cost of ownership includes costs associated with the initial investment, that is, interest on the investment and its repayment over a specified number of years related to its lifetime. The sum of these is usually taken as a fixed percentage of investment each year; for example, for a 20 year amortization and 8% interest rate, the annual cost is 0.10185 of the investment.

Operating costs are primarily for power requirements for pumping water and moving air in the system, summed over the yearly operating time of the system. Maintenance costs include repairs, replacement of glass in collectors, or any other costs of keeping the system in operating condition. Consideration of these costs leads to the conclusion that maintenance must be minimized if solar heating is to be economically viable, particularly when labor costs are to be charged as part of the maintenance expense.

The annual cost of delivering solar heat, that is, of owning, operating, and maintaining the solar heating system, in dollars per year, can be formulated as follows:

$$C_{S,a} = (C_C A_C + C_{ST} + C_E)I + P C_P + C_{MM} + C_{ML} \quad (12.5.1)$$

where $C_{S,a}$ = the annual cost of the solar energy system;
C_C = the capital cost per unit area of collector (a function of design, number of covers, mounting, and so on;
A_C = the area of collector;
C_{ST} = the capital cost of storage (medium, container, and insulation);
C_E = the capital cost of equipment, pumps, piping, ducts controls, and so on;
I = fraction of investment to be charged per year, interest, and depreciation;
P = the annual power requirements of the solar energy system;
C_P = the unit cost of power;

C_{MM} = the annual cost of maintenance materials; and
C_{ML} = the annual cost of maintenance labor.

The first term represents costs of owning the system, and is usually the largest term in the equation. The second term covers power costs and similar items, and is usually small in well-designed systems, particularly water systems, although it cannot be assumed to be negligible without calculating it. The maintenance terms are low for well-engineered systems, but may be dominant and deterrent factors in systems that require more than nominal maintenance during their lifetime.

There can be other kinds of costs chargeable to the solar energy system. For example, insurance or real estate taxes (if imposed on assessed valuation of solar heating equipment) may represent an additional significant increment in annual cost. On the other hand, there is pending tax legislation in several states which may provide tax relief or incentives for use of solar heating.

The cost per unit of energy delivered by the solar system can be obtained by dividing the annual cost of the system, $C_{S,a}$ [from Eq. (12.5.1)] by the integrated solar energy delivered to the building over a year. This is obtained from the analysis of thermal performance and it depends on many factors, the most important of which is the balance between the energy needs of the building and the capacity of the collector. Because of the stochastic nature of the distribution of heating loads over time, a large collector on a given building will be oversized for that building longer than a small collector would be, and the cost per unit of delivered solar energy rises as collector size increases. On the other hand, for small collector sizes, costs (C_E) that are essentially independent of collector area, dominate. Under these circumstances there can be a least-cost solar energy system, that is, a collector-building combination that leads to a minimum cost of delivered solar energy.

The next step in a cost analysis is to enter the cost of auxiliary (i.e., conventional) energy. This can be formulated in a manner similar to that of the solar system, but the dominant costs will be for fuel.

$$C_{A,a} = C_B I' + Q_A C_F + P'C_P + C'_{MM} + C'_{ML} \qquad (12.5.2)$$

where $C_{A,a}$ = the annual cost of supplying auxiliary or conventional energy;
C_B = the first cost of furnace and associated equipment;
I' = the fraction of investment in this equipment to be charged per year (may be different from I);

Q_A = the annual energy to be supplied from auxiliary (fuel) source;

C_F = the cost of fuel (or electricity if used for heat) per unit of energy delivered (that may depend on Q_A);

P' = the annual power requirement of the auxiliary system;

C_P = the unit cost of power, as in Eq. (11.5.1);

C'_{MM} = the annual cost of maintenance materials for the auxiliary system; and

C'_{ML} = the annual cost of maintenance labor for the auxiliary system.

The total annual cost of the solar energy system with auxiliary is then the sum of the solar and auxiliary costs:

$$C_{T,a} = C_{S,a} + C_{A,a} \qquad (12.5.3)$$

The analysis of the yearly thermal performance of a system determines how much solar energy it can deliver and how much must be provided by the auxiliary source. When this information is available for the system and for the location in question, annual total cost can be calculated and the effects of collector area (or other design variables) on this cost can be assessed.

It is now possible to compare solar energy systems (including auxiliary) and conventional systems. Equation (12.5.2) gives the cost of the conventional system and (12.5.3) the solar system. This comparison involves considering the cost of owning the solar system and purchasing some fuel on the one hand, and pruchasing a larger quantity of fuel on the other. If the annual cost of the solar system, including the cost of auxiliary, is less than that of the conventional system, then there is direct economic advantage to the use of the solar system.

The next question concerns the optimum (i.e., least total annual cost) mix of solar energy and auxiliary. All of the factors that affect thermal performance of the solar heating system will in turn affect the optimum combination. However, for purposes of discussion, consider the collector area as the primary design parameter of interest. The larger the area, the larger the fraction of total annual heating loads will be met by solar energy, and the less fuel will be required. Assuming that the costs of owning and maintaining the conventional source are fixed regardless of collector size and thus do not affect the optimum, and that a well-designed solar system has negligible maintenance costs, a simplified form of Eq. (12.5.3) can be used to determine the optima:

$$C_{T,a} = (C_C A_C + C_{ST} + C_E)I + PC_P + Q_A C_F + P'C_P \quad (12.5.4)$$

The power costs are usually small, although they may be significant, particularly with air systems. Thus the two dominant terms are the first, which increases as collector area increases, and the third (the auxiliary energy cost), which decreases as collector area increases. This analysis may require modification, for example, if the nature of the auxiliary energy supply to be used is in any way a function of the amount of auxiliary required. Most probably, auxiliary energy will have to be supplied from a source stored on site, since a large number of solar heating systems simultaneously calling on a utility for auxiliary energy would place intolerable short-time loads on that utility. Also, the auxiliary heating capacity for a building must be able to meet the worst possible conditions that can be encountered.

A major economic study of solar house heating, based on these considerations, has been reported by Tybout and Löf (1970), and Löf and Tybout (1973). They devised a thermal model for a solar heating system to compute annual thermal performance (based on one year's meteorological data), and they developed a set of cost assumptions to calculate costs of delivered solar energy for houses of two sizes in each of eight U. S. locations of differing climate types (Miami, Albuquerque, Phoenix, Santa Maria (California), Charleston, Seattle, Omaha, and Boston). Several system design parameters were studied in addition to collector area, to establish the range of optimum values, including angle of tilt of collector toward equator, number of covers, and heat storage capacity per unit collector area. Their results (which are in general agreement with the conclusions of other authors), can be summarized as follows:

1. The optimum tilt is in the range of the latitude plus 10° to the latitude plus 20°, and variation of 10° either way outside of this range, that is, from latitude to latitude plus 30°, has relatively little effect on the cost of delivered energy for heating.

2. The best number of (ordinary) glass covers was found to be two for all locations except those in the warmest and least severe climates, that is, Miami and Phoenix, where one cover produces less expensive energy from the solar heating system.

3. Storage capacity per unit collector area was indicated to be in the range of 200 to 300 kJ/m² °C, or for the water storage systems studied, 50 to 75 kg/m². Increasing the size

of storage for fixed collector designs had relatively small effect on the cost of delivered solar energy or on the fraction
of total heating loads carried by solar.

Using these optimum ranges of these design parameters, Tybout
and Löf then considered the major design parameter, collector
size, or the fraction of the total annual heating load to be
carried by solar energy, for particular houses in particular
locations. The results of a very large number of computations
are summarized in Figures 12.5.1 and 12.5.2, which show, based
on their cost assumptions, the variation in cost of delivered
solar energy as a function of the percent of total heat load
supplied by solar energy. At the low percentages (small collectors), the costs of those items of equipment that are independent of collector area dominate. For very large collectors the

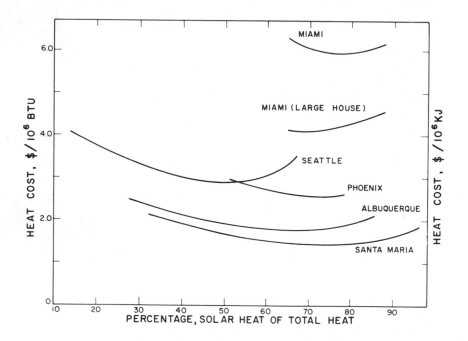

Figure 12.5.1   Influence of collector size on cost of solar
heat. All of the houses have 15,000 Btu/degree-day loss rates,
except for the "large" Miami house which has a loss rate of
25,000 Btu/degree-day [adapted from Tybout and Löf (1970) and
Löf and Tybout (1973)].

energy that can be delivered and used from increments in collector area is diminished, the average energy gain per unit area of collector drops, and the cost of that energy rises.

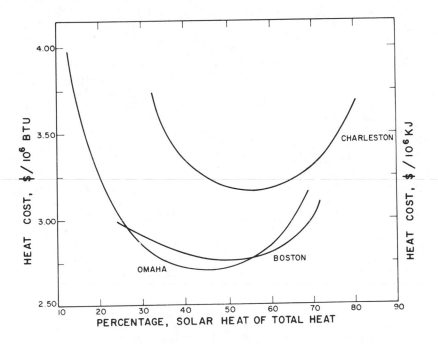

Figure 12.5.2 Influence of collector size on cost of solar heat (15,000 Btu/degree-day houses) [adapted from Tybout and Löf (1970) and Löf and Tybout (1973)].

Tybout and Löf developed and used costs including the following: collectors, $2 or $4/ft ; equipment costs, $375; interest rate, 6%; ammortization period, 20 yrs. Other sets of cost data, other system designs, house designs, and climates, may lead to different optimum fractions of energy to be delivered by solar energy. The minimum costs may be higher or lower. The results, however, can be shown by cost curves similar to those of Tybout and Löf.

12.6 SOME ARCHITECTURAL CONSIDERATIONS

Optimization studies of solar heating systems that have been done to date suggest that a collector area in the range of one-third to two-thirds of the floor area of the house will result in least-cost systems (with the result in any specific case very much a function of the design of the building, the system, and the climate). The architect faces a new set of problems and constraints in designing into or onto the building a collector of this size, relative to the building and with appropriate orientation. Space must be provided in the structure for energy storage units, piping and ducts, controls, auxiliaries, and all associated equipment. Anderson et al. (1955) have addressed these and related questions on solar house architecture. Although most of the work to date has been on single-family residential buildings, similar considerations will apply to institutional buildings.

The collector may be a part of the envelope of the building (as in MIT House IV) or separate (as in the Denver Solar House). The orientation of the collector is substantially fixed and if it is part of the envelope of the house, the collector will probably become an important or dominant architectural feature of the structure. It may be possible for the collector to serve as part of the weatherproof enclosure and thus allow a reduction in cost of roofing or siding; such a reduction is actually a credit to reduce the cost charged to the collector. Separate collectors, on the other hand, can permit greater flexibility in house design and allow buildings that are more conventional (contemporary or traditional) in appearance.

Storage is usually not a major architectural problem, other than recognizing the need for appropriate space and access within the structure. The volume of storage, per unit area of collector, depends on the system used. Pebble bed storage units will have a volume of roughly 0.17 to 0.26 m^3/m^2 of collector and water tanks about 0.050 to 0.075 m^3/m^2. The volume of heat of fusion storage systems would be less than that of water systems. The most common location for storage in buildings constructed so far is in basement space.

Providing solar heat to larger buildings, such as apartment buildings, presents a special set of problems. It may be necessary to consider vertical mounting for collectors, thus causing a significant reduction in their performance. This possibility is being studied by Lorsch and Niyogi (1971). Otherwise, collectors may be mounted like awnings, with improvements in performance but with increased cost of installation. If the building geometry is such that the roof area is adequate, banks of collectors can be mounted on the roof with appropriate piping or ducts leading to the spaces to be heated.

Essentially, all solar heating studies to date have been concerned primarily with new buildings designed to include solar heating systems. Adding solar heating systems to most existing buildings presents a more formidable task, and consideration must be given in the future to the problems of designing effective add-on solar heating systems and the attendant architectural problems.

Figure 12.6.1 presents photographs of the Denver house, MIT House IV, and a new experimental house constructed by Colorado State University at Fort Collins, Colorado, in 1974. Two other houses are pictured in Chapter 14.

12.7 MODELING OF SOLAR HEATING SYSTEMS

Although the basic technology of solar heating is available, we do not yet have a generally satisfactory design procedure. A method in use at the present time is to model the system components and the building and simulate the operation using the model in the desired climate as outlined in Chapter 10. By going through the simulation for several collector areas (and for appropriate ranges of values of other design parameters as may be necessary), and in terms of costs of the system, the design leading to least-cost delivered energy can be determined. An example of this procedure is given in the following section.

A necessary step in the procedure is to model the thermal performance of the building. Several approaches to this problem are possible, depending on the type of building and the degree of detail wanted. The simplest approach is to use a "degree-day" model, in which the heating loads are taken as a function of the difference between a fixed "effective indoor" temperature (usually 18°C) and the ambient temperature. At the other extreme, detailed analyses can be based on methods outlined in the *ASHRAE Handbook of Fundamentals* (1972). Computer programs are available (e.g., from the National Bureau of Standards) to carry out these analyses. An alternative approach, used by Butz and noted in the following section, is to write simpler programs that do not model the building in detail but that do take into account the thermal capacitance of the structure, windows, solar radiation, wind speed, and energy generation in the building.

Modeling of solar heating systems and simulation of their operation are tools that now require detailed meteorological information (e.g., hourly values of solar radiation, wind speed, and dry bulb temperature), and may use a significant level of

Figure 12.6.1 Solar heated buildings. Top, the Denver house. Middle, MIT House IV. Bottom, the Colorado State University house.

computing time. In its present form, simulation may be used for
house system design, but it will in the future probably be restricted to larger institutional buildings where higher levels
of engineering effort are put on a single building design. There
is also the problem that meteorological data, as detailed as is
needed, are available for only a few locations. The development
of simplified design procedures for routine engineering use,
which will take into account the weather, its variability, and
the transient performance of the heating system, is a problem
not yet solved.

12.8 PERFORMANCE AND COSTS CALCULATIONS OF A SOLAR HEATING SYSTEM

A recent study by Butz (1973) and Butz et al. (1974) uses modeling methods to estimate the thermal performance of a building
equipped for solar service hot water heating, space heating,
and air conditioning. In this chapter we discuss the hot water
and heating functions, their modeling, and computed performance.
The computed thermal performance serves as a basis for an economic analysis. In the next chapter the performance and economics of the air conditioning part of the system are added. Although this system is not optimum in all respects, and although
the models of one or more of the components are subjects for
refinement and improvement, the study shows the manner in which
models are used to predict thermal performance and how the performance predictions can be used to arrive at minimum total cost
systems. (Note: Butz's cost analysis was based on the combination of solar heating, hot water, and cooling. Here we have
separated the heating and hot water analysis and based a cost
study on his system without cooling.)

The system (without the air conditioner) is shown schematically in Figure 12.8.1. Most of the design parameters were
selected, based on Löf and Tybout (1973) and others, for the
Albuquerque, New Mexico, climate; performance calculations were
based on a year's observations of hourly meteorological data for
Albuquerque. Table 12.8.1 lists the major design parameters assumed by Butz. The principle variable design parameter was collector area, and the discussion below leads to determination of
the optimum collector area for the "Albuquerque house."

The major components of the system, and the way in which
they were modeled, can be summarized as follows:

A. The solar collector is a flat-plate water heater having
characteristics noted in the table. Its heat capacity is neglected, and it is modeled by Eq. (7.7.5). The collector and

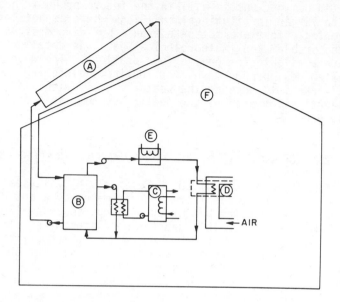

Figure 12.8.1 Schematic of solar heating, and hot water system
[from Butz (1973)]. A, collector. B, storage tank. C, hot
water system. D, space heating heat exchanger. E, main auxiliary heater. F, the building.

TABLE 12.8.1 Major Design Parameters of the Solar Heating System as Modeled by Butz.

Mass of water storage = 61.0 kg/m^2

Area of collector = 13.9, 32.5, 60.4, and 88.3 m^2

Collector F_R = 0.878

Collector U_L = 4.56 W/m^2 °C

Number of covers = 2

Slope of collector from horizontal = 40°

Floor area of house = 167 m^2

Latitude = 35°N

Maximum allowable collector outlet temperature = 110 °C

---

the balance of the system are assumed to operate at a maximum pressure of two atmospheres absolute, so the maximum collector temperature is limited to 110 °C. Although not designed into the system explicitly, the assumption is made that a means is provided of rejecting excess energy to keep the collector exit fluid temperature below this level.

  B. The storage tank is a water tank having fixed size in relation to the collector area of 61.0 kg/m^2. It was modeled according to the method of Chapter 9 as both a three-segment, partially stratified tank by equations similar to those in Chapter 9, and as a fully mixed tank by Eq. (9.3.2). The results noted here are based on the (conservative) assumption of no stratification. Thermal losses from the tank are to the interior of the building.

  C. The service hot water system is composed of several elements, based on the idea that a conventional gas or electric water heater might be used with the addition of solar energy as a means of energy supply. Solar energy is provided by a heat exchanger between the main storage tank and the service hot water tank. Hot water auxiliary is separate from the main auxiliary and is added directly to the tank. The tank is modeled as fully mixed with water at a uniform temperature, and auxiliary energy added to the tank whenever the tank temperature cannot be maintained at a minimum control temperature by energy transferred across the heat exchanger from the main storage tank.

  D. A water-to-air heat exchanger is the means by which energy from the tank is transferred to the building. It is modeled by the effectiveness-NTU method.

  E. The main auxiliary energy supply (the furnace) is a two stage heater in series with the storage tank and water-to-air heat exchanger. (Under the design conditions used, this arrangement can, under some circumstances, result in the energy addition by the auxiliary heater being larger than the load on the heat exchanger, thus adding auxiliary energy to the storage tank. This renders the collector less efficient that it otherwise would be.)

  F. The building was assumed to be of 167 m^2 floor area and had a heating requirement of approximately 32,300 kJ/degree-°C-day (17,000 Btu/degree-day). The walls are assumed to be of standard construction, with good insulation, and the collector was assumed to be mounted so that it would not affect the energy balance of the house. The building was modeled by a thermal network to compute thermal gains and losses and heat capacity effects of the structure. Each of the four walls and the roof was modeled as having three nodes, with capacitance and conductances representative of standard building construction. The

walls were assumed to have 15% window area. The interior structure and furnishings were represented by single node; the temperature of this interior node was the temperature on which control was based. Infiltration, and energy and humidity generation inside the building were included.

G. Collector controls, not shown in the diagram, are based
on the temperature in the storage tank (in the bottom of the
tank, in the case of stratified storage) and the computed collector output temperature as if the collector were operating.
Whenever the latter is higher, that is, when useful energy can
be gained, the collector is operated by turning on the pump to
circulate water. This very closely resembles a control based
on measurement of the top header temperature and storage tank
temperature in an actual system.

H. Controls on the heating system are not shown in the
diagram. The primary variable on which control is based is the
computed temperature of the interior node of the building (the
room temperature). Heating control is based on a set of four
control temperatures; the operation of this control system is
indicated in Figure 12.8.2.

TEMPERATURE / Heating		High Auxiliary	Low Auxiliary	Space Heater
T_{low}	↑	Off	Off	Turns Off[*]
$T_{hot,0}$	↑	Off	Turns Off[*]	On
	↓	Off	Off	Turns On
$T_{hot,1}$	↑	Turns Off[*]	On	On
	↓	Off	Turns On	On
$T_{hot,2}$	↓	Turns On	On	On

[*] Turns off if on

Figure 12.8.2 Schematic of the type of controls and modes of
operation of the heating system modeled for the Albuquerque
house. ↑ means as room temperature rises past the control temperature; ↓ means as room temperature drops past the control
temperature.

For each of the four collector areas, this system was
"operated" in the Albuquerque climate, to produce heating when
needed and service hot water for the whole year. An example of
the daily averages for each month of the major energy quantities
in the system, for a collector area of 32.5 m², is shown in
Figure 12.8.3. Computed integrated energy quantities for months

for two of the collector areas are shown in Tables 12.8.2 and
12.8.3. Figure 12.8.4 shows a cumulative plot of energy flows
for the building for a collector area of 32.5 m², in a manner
analogous to that shown by Löf et al. for the Denver house, and
Engebretson for the MIT House IV. For purposes of this plot,
the year starts at the beginning of the heating season (the beginning of September).

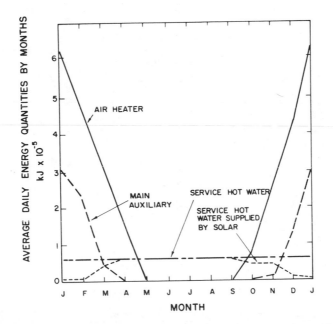

Figure 12.8.3 Computed monthly variations of average daily
energy quantities for the Albuquerque House, for a collector
area of 32.5 m² [adapted from Butz (1973)].

     Figure 12.8.5 shows other kinds of information that can be
derived from the thermal analysis; it indicates annual energy gain per unit of collector area as well as efficiency, both
of which are functions of collector size. As the collector is
made larger, the fraction of the year during which incremental
additions to the collector can be used drops substantially,
leading to diminishing use factor on the total collector. In
other words, the larger the system, the greater the fraction
of the time it is oversized. Also, smaller systems operate at
lower temperature and thus the efficiency of collection is higher.

TABLE 12.8.2 Computed Monthly Performance of the Albuquerque House.*

Month	Heating			Hot Water			Total		Q_{SOL}
	Total	Solar	Aux.	Total	Solar	Aux.	Solar	Aux.	
Jan.	18.7	9.6	9.1	2.0	0.3	1.7	9.9	10.8	28.6
Feb.	12.7	6.6	6.1	1.8	0.4	1.4	7.0	7.5	21.2
Mar.	8.6	7.7	0.9	2.0	1.4	0.6	9.1	1.5	32.4
Apr.	4.5	4.5	0.0	1.9	1.9	0.	6.4	0.0	25.3
May	0	0	0	2.0	2.0	0	2.0	0	24.3
June	0	0	0	1.9	1.9	0	1.9	0	24.1
July	0	0	0	2.0	2.0	0	2.0	0	23.8
Aug.	0	0	0	2.0	2.0	0	2.0	0	23.4
Sept.	0	0	0	1.9	1.9	0	1.9	0	27.5
Oct.	2.0	2.0	0.0	2.0	1.4	0.6	3.4	0.6	23.8
Nov.	6.9	6.7	0.2	1.9	1.4	0.5	8.1	0.7	28.3
Dec.	12.1	8.3	3.8	2.0	0.6	1.4	8.9	5.2	23.6
	65.5	45.4	20.1	23.4	17.2	6.2	62.6	26.3	306.3

*Collector area = 32.5 m², energy quantities = kJ × 10⁻⁶.
Annual η = 62.6/306.3 = 0.20; % Total annual load carried by solar = 70%.

TABLE 12.8.3 Computed Monthly Performance of the Albuquerque House.*

Month	Heating			Hot Water			Total		Q_{SOL}
	Total	Solar	Aux.	Total	Solar	Aux.	Solar	Aux.	
Jan.	18.7	15.0	3.7	2.0	0.8	1.2	15.8	4.9	53.3
Feb.	12.7	9.6	3.1	1.8	1.0	0.8	10.6	3.9	39.3
Mar.	8.6	8.6	0	2.0	1.8	0.2	10.4	0.2	60.1
Apr.	4.5	4.5	0	1.9	1.9	0	6.4	0	46.9
May	0	0	0	2.0	2.0	0	2.0	0	44.2
June	0	0	0	1.9	1.9	0	1.9	0	44.6
July	0	0	0	2.0	2.0	0	2.0	0	44.2
Aug.	0	0	0	2.0	2.0	0	2.0	0	43.5
Sept.	0	0	0	1.9	1.9	0	1.9	0	51.0

continued

TABLE 12.8.3   (Continued)

Month	Heating			Hot Water			Total		Q_{SOL}
	Total	Solar	Aux.	Total	Solar	Aux.	Solar	Aux.	
Oct.	2.0	2.0	0	2.0	1.7	0.3	3.7	0.3	44.2
Nov.	6.9	6.9	0	1.9	1.9	0	8.8	0	52.5
Dec.	12.1	11.1	1.0	2.0	1.0	1.0	12.1	2.0	43.9
	65.5	57.7	7.8	23.4	19.9	3.5	77.6	11.3	567.7

*Collector area = 60.4 m², energy quantities = kJ × 10⁻⁶.
Annual η = 0.13; % Total annual load carried by solar = 87%.

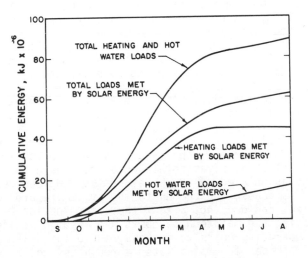

Figure 12.8.4 Computed cumulative solar energy inputs and total
loads for the Albuquerque House, for a collector area of 32.5 m².
Data are from Table 12.8.2.

With these thermal performance estimates, we turn to determination of the collector area leading to least cost for heating
and hot water. In this analysis, total costs are estimated,
based on assumed costs of energy from conventional sources (i.e.,
on cost of fuel, C_F, in dollars/10⁶ kJ) and on assumed costs per

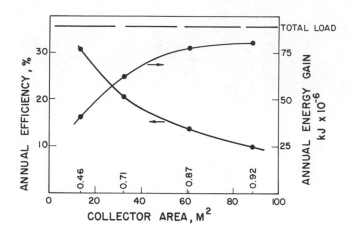

Figure 12.8.5 Annual energy gain and annual efficiency of solar
energy collection, for heating and service hot water, as a function of collector area for the Albuquerque House. Total energy
requirements for this house are 88.5×10^6 kJ for heating and
hot water for this year. Numbers below data points show fraction of the total annual load carried by solar energy.

unit area of collector (i.e., first cost, C_C, and annual charge,
I). Alternatively, actual costs of collector and associated
equipment and of fuels can be used for the location, building,
and time in question. These "actual" costs, however, are variable with time, and here C_C and C_F are taken as parameters.
     Equation (12.5.4) gives total annual cost as a function of
collector and associated costs, power costs, and fuel costs.
Furnace cost is considered common to the system regardless of
collector size, and does not affect the determination of best
area. Operating and maintenance costs are assumed negligible;
incremental operating costs to be assigned to the solar system
are primarily those of pumping water through the collector system, which are small in a well-designed liquid heating system.
Thus, Eq. (12.5.4) reduces to a relative annual cost, a cost
above base costs, $C_{T,a}$, of

$$C_{T,a} = (C_C A_C + C_{ST} + C_E)I + Q_A C_F \qquad (12.8.1)$$

If storage capacity is directly related to collector area,

$$C_{T,a} = [(C_C + C'_{ST})A_C + C_E]I + Q_A C_F \qquad (12.8.2)$$

where c'_{ST} = cost of storage per unit area of collector. The following assumptions relating to costs were made:

I = 0.10185, corresponding to an interest rate of 8% and 20-year life;
c'_{ST} = \$8.00/m^2, based on a tank cost of \$0.132 per kilogram capacity and 61.0 kg/m^2, storage capacity per unit area of collector; and
C_E = \$250, for additional piping, pumps, controls modifications to service hot water system, and so on, independent of collector area.

Parameters in this cost analysis are C_C, which is the cost of collector, taken as 20, 40, and \$60/m^2, C_F, which is the cost of auxiliary energy, taken as 2, 4, \$6/10^6 kJ of delivered energy. The cost above base is then

$$c_{T,a} = [(C_C + 8.00)A_C + 250]0.10185 + Q_A C_F \quad (12.8.3)$$

where C_C and C_F are cost parameters and A_C and Q_A are area of collector and yearly auxiliary energy required, determined from the thermal analysis.

The annual cost above base cost is plotted as a function of collector area in Figure 12.8.6, for three values of collector cost and for three values of fuel cost. In each case, except those of lowest fuel cost and highest collector cost, the cost curves show minimum values; the best collector area increases as the cost of energy from fuel rises, and as collector cost decreases. The optimum collector areas are, in this example, in the range of 20 to 50 m^2; at these areas the fractions of total energy needs supplied by solar are in the range of one half to three quarters. (See Figure 12.8.5.)

The horizontal lines on Figure 12.8.6 that are marked with values of C_F are costs of fuel if the house were heated entirely from the conventional source. Wherever the curves for the combined system lie below these lines, the annual cost of the combined system will be less than that of the conventional system. For example, at a fuel cost of \$4/10^6 kJ, combined systems with collector costs of \$40/m^2 will be less expensive in the collector area range of 5 to 58 m^2; if collector costs are \$60/m^2, combined systems in a collector area range of near zero to 33 m^2 show lower costs than the conventional system.

This example, while not based on a system optimized in all respects, illustrates how models can be used to estimate thermal performance, and how thermal performance can be related to costs in assessing the best combinations of solar and fuel from

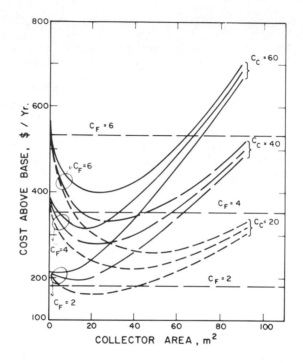

Figure 12.8.6. Annual cost (above the cost of furnace and piping common to both solar and conventional systems) of solar and
fuel heating of a 167 m² house in Albuquerque, for collectors
costs C_C of 20, 40, and $60/m²$, as a function of collector area.
For each collector cost, curves are shown for fuel costs C_F of
2, 4, and $6/10^6$ kJ delivered.

least-cost considerations. In Chapter 13, this same system will
be considered again, with the addition of summer operation of
an absorption air conditioner. The numbers obtained in this
example are specific to the design and climate used in the simulation, and should not be applied to other designs or locations.

REFERENCES

Anderson, L. B., Hottel, H. C., and Whillier, A., Solar Energy
Research, 47, Madison, University of Wisconsin Press (1955).
"Solar Heating Design Problems."

ASHRAE Handbook of Fundamentals, New York, Am. Soc. Heating, Refrigerating and Air Conditioning Engineers, 1972.

Bliss, R. W., Air Conditioning, Heating and Ventilating, *92*, (October 1955). "Design and Performance of the Nation's Only Fully Solar-Heated House." See also Proceedings of the World Symposium on Applied Solar Energy, 151, SRI, Menlo Park, California (1956).

Böer, K., Paper presented at International Solar Energy Society Conference, Paris (1973). "A Combined Solar Thermal Electrical House System."

Butz, L. W., M.S. Thesis in Mechanical Engineering, Madison, University of Wisconsin (1973). "Use of Solar Energy for Residential Heating and Cooling."

Butz, L. W., Beckman, W. A., and Duffie, J. A., Solar Energy, *16*, 129 (1974). "Simulation of a Solar Heating and Cooling System."

Close, D. J., Dunkle, .R V., and Robeson, K. A., Mech. Chem. Engr. Trans., Inst. Engrs. Australia, *MC4*, 45 (1968). "Design and Performance of a Thermal Storage Air Conditioner System."

Engebretson, C. D. and Ashar, N. G., Paper 60-WA-88 presented at the New York ASME Meeting (1960). "Progress in Space Heating with Solar Energy."

Engebretson, C.D., Proceedings of the UN Conference on New Sources of Energy, *5*, 159 (1964). "The Use of Solar Energy for Space Heating - M.I.T. Solar House IV."

Hottel, H. C. and Whillier, A., Transactions of the Conference on the Use of Solar Energy, *2*, 74, University of Arizona Press, 1958. "Evaluation of Flat-Plate Solar Collector Performance."

Löf, G. O. G., El-Wakil, M. M., and Chiou, J. P., Trans. ASHRAE, *77* (October 1963). "Residential Heating with Solar Heated Air —The Colorado Solar House."

Löf, G. O. G., El-Wakil, M. M., and Chiou, J. P., Proceedings of the UN Conference on New Sources of Energy, *5*, 185 (1964). "Design and Performance of Domestic Heating System Employing Solar Heated Air—The Colorado House."

Lof, G. O. G., Proceedings of the UN Conference on New Source of Energy, *5*, 114 (1964). "Use of Solar Energy for Heating Purposes: Space Heating."

Löf, G. O. G. and Tybout, R. A., Solar Energy, *14*, 253 (1973). "Cost of House Heating with Solar Energy."

Lorsch, H. G. and Niyogi, B., Report NSF/RANN/SE/GI 27976/TR72/18, to NSF from University of Pennsylvania (August 1971). "Influence of Azimuthal Orientation on Collectible Energy in Vertical Solar Collector Building Walls."

Space Heating with Solar Energy, R. W. Hamilton, ed., Cambridge, Massachusetts Institute of Technology, 1954.

Telkes, M. and Raymond, E., Heating and Ventilating, *80* (1949). "Storing Solar Heat in Chemicals—A Report on the Dover House."

Thomason, H. E. and Thomason, H. J. L., Solar Energy, *15*, 27 (1973). "Solar Houses/Heating and Cooling Progress Report."

Tybout, R. A. and Löf, G. O. G., Natural Resources J., *10*, 268 (1970). "Solar Energy Heating."

13. SOLAR COOLING

The use of solar energy to drive cooling cycles has been considered for two different but related purposes. The first of
these is to provide refrigeration for food preservation, and
the second to provide comfort cooling. In section 13.1 we briefly review some of the literature relating to both of the applications, since there is a common underlying technology. From
then on, we consider problems relating to solar air conditioning. In particular, for application in the United States, we
address questions of the use of flat-plate collectors for both
winter heating and summer cooling.

There is less experience with solar cooling than solar
heating. Several solar heated buildings have been designed,
built, operated for extended periods, carefully studied, and the
results published. In contrast, solar cooling has been the object of few short-time experiments that indicate the technical
feasibility of solar operation of absorption air conditioners,
but that have not served to evaluate systems or pinpoint critical design factors. So, solar cooling is in an earlier stage
of development than solar heating. In the last several years,
modeling of cooling systems has been accomplished in a manner
that does allow a start on evaluations and identification of
cirtical design areas.

In this chapter we concentrate on operation of air conditioning equipment with the same flat-plate collectors that are
used for winter heating. In principle, it is possible to convert solar to electrical or mechanical energy, and use that
energy to drive a mechanical refrigeration system. The major
problem lies in the first conversion step; if that can be accomplished economically, the way will be open to solar operation
of compression-type air conditioners that will avoid some of the
problems of absorption systems, and a new Chapter 13 will then
have to be written.

Chapter 14 notes some other methods of solar heating and
cooling to illustrate the diversity of approaches that have
been proposed to these problems.

13.1 *REVIEW OF SOLAR ABSORPTION COOLING*

Two approaches have been taken to solar operation of absorption coolers. The first is to use continuous coolers, similar in construction and operation to conventional gas or steam fired units, and with energy supplied to the generator from the solar collector-storage-auxiliary system whenever conditions in the building dictate the need for cooling. The second approach is to use intermittent coolers similar in concept to that of commercially manufactured food coolers used many years ago in rural areas (the Crosley "Icyball") before electrification and mechanical refrigeration were widespread. Intermittent coolers have not, to our knowledge, been used for air-conditioning, nor have they been carefully studied from the standpoint of their possible applicability to solar air conditioning.

Experience to date suggests that continuous absorption cycles may be adapted to operation from flat-plate collectors. A schematic diagram of one possible arrangement of such a system is shown in Figure 13.1.1. The present temperature limitations on operation of flat-plate collectors restricts consideration among commercial machines to lithium bromide-water systems. LiBr-H_2O machines require cooling water for cooling the absorber and condenser, and their use will probably require use of a cooling tower. Solar operation of ammonia-water coolers such as those now marketed on a large scale is difficult because of the high temperatures required in the generator.

Little effort has been spent to date on design of absorption coolers for operation with flat-plate collectors. The use of coolers specifically designed for operation from solar energy could probably result in lower generator operating temperatures, better operation over a range of temperature, and energy input levels to the generator from a collector. If cooling requirements rather than heating loads fix collector size requirements, it may be an advantage to design coolers with higher than usual coefficient of performance (COP). For example double-effect evaporators can be used to decrease energy input requirements [see Whitlow and Swearingen (1959) or Chinnappa (1973)]. In other words, the conditions and constraints of solar operation may lead to cooler designs different than those for fuel operation.

Continuous abosrption cooling by solar operation represents a technically feasible approach to solar cooling. A commercial lithium bromide-water air conditioner, slightly modified to allow supplying the generator with hot water rather than steam, was operated from a flat-plate water heater at the University of Wisconsin [Chung, et al. (1963)]. The performance of the experimental system was measured, both on an instantaneous basis and

and with the performance integrated over "full" days of operation; the results are summarized in the next section.

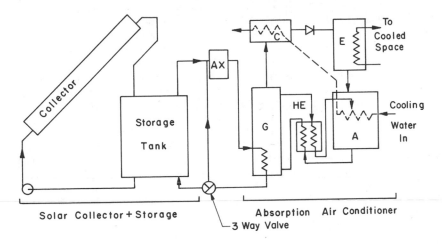

Figure 13.1.1 Schematic of a solar operated absorption air conditioner. *AX* is auxiliary energy source. The essential components of the cooler are: *G*, generator; *C*, condenser; *E*, evaporator; *A*, absorber; *HE*, heat exchanger to recover sensible heat.

        An analytical study of solar operation of a LiBr-H_2O cool-
er and flat-plate collector combination has been carried out
[see Duffie and Sheridan (1965)] to identify critical design
parameters and to assess the effects of operating conditions on
integrated solar operation. Under the assumptions made in this
study, design of the sensible heat exchanger between absorber
and generator, cooling water temperature, and generator design
are important; the latter is more critical here than in fuel-
fired coolers because of the coupled performance of the collector and cooler. An experimental program was also developed at
Queensland, Australia, in a specially designed laboratory house
[see Sheridan (1970)].
        Chinnappa (1967) has calculated the performance of a con-
tinuous LiBr-H_2O cooler with solar operation for the climate of
Colombo, Ceylon, noted the possibility of refrigerant-absorbent
storage for energy storage, and made an initial cost estimation
that indicated that the solar operation would be competitive
with electrically operated compression systems in Colombo.
        Lauck et al. (1965) simulated a particular solar and gas
operated heating and absorption cooling system, using selected

winter and summer days' meteorological data for Madison, Wisconsin. The computer used at that time was slow; the conclusion reached was that simulation is a practical guide to experiment and this appears to be truer today than it was 10 years ago.

Farber et al. (1960,1970) have studied a series of solar-operated ammonia-water coolers, using flat-plate collectors, without storage. The temperature range of water supply to the generator was generally about 60 to 93 °C; the condensing water temperature was not specified. Ammonia concentrations in the absorber and generator were typically 58% and 39%. Operation was continuous and at varying rates, depending on energy supply.

Thus, there have been some significant experimental and analytical studies of this class of systems, but the scope of the individual studies has been limited. The studies have not adequately treated the problems of energy storage, and only one has dealt with questions of combined heating and cooling systems.

Intermittent absorption cooling may be an alternative to continuous systems. The limited work to date on these cycles has been largely directed at food preservation rather than comfort cooling. These early efforts are germane to the air conditioning problem, however, because these cycles offer potential solutions to the energy storage problem. In these cycles, distillation of refrigerant from the absorbent occurs during the regeneration stage of operation, and the refrigerant is condensed and stored. During the cooling portion of the cycle, the refrigerant is evaporated and reabsorbed. A schematic of the simplest of these processes is shown in Figure 13.1.2. Thus, "storage" is in the form of separated refrigerant and absorbent. Modifications of this basically simple cycle, using pairs of evaporators, condensers, or other arrangements, might result in an essentially continuous cooling capacity and improved performance.

Refrigerant-absorbent systems used in intermittent cycles have been NH_3-H_2O and NH_3-NaSCN. In the latter system, the absorbent is a solution of NaSCN in NH_3, with NH_3 the refrigerant. This system has been studied by Blytas and Daniels (1962) and by Sargent and Beckman (1968), and it appears to have good thermodynamic properties for cycles for ice manufacture. Williams et al. (1958) reported an experimental study of an intermittent NH_3-H_2O cooler using a focusing collector for regeneration.

Chinnappa (1961,1962) and Swartman and Swaminathan (1970) have been experimentally studying the operation of intermittent NH_3-H_2O machines in which flat-plate collectors have served as the energy supply. The absorber and generator are separate vessels. The generator is an integral part of the flat-plate collector, with refrigerant-absorbent solution in the tubes of the collector circulated by a combination of thermosyphon and bubble

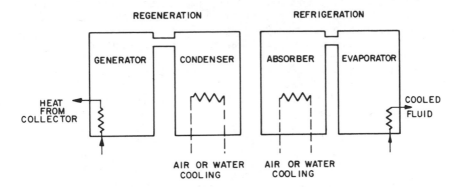

Figure 13.1.2 Schematic of an intermittent absorption cooling cycle. On left, the regeneration cycle. On the right, the refrigeration cycle. The generator-absorber is a single vessel performing both functions, and the condenser-evaporator is also a single vessel performing both functions.

pump action. Using approximately equal cycle times for the regeneration and refrigeration steps (5 to 6 hrs each), overall coefficients of performance were found to be about 0.06 at generator temperatures rising from ambient to about 99 °C during regeneration. Evaporator temperatures were below 0 °C. With cooling water available at about 30 °C, the effective cooling per unit area of collector surface per day for the experimental machine was in the range of 50 to 85 kJ/m^2 for clear days.

13.2 PERFORMANCE OF A SOLAR OPERATED ABSORPTION COOLER

Limited experiments have been done on solar operation of continuous LiBr-H$_2$O absorption coolers with flat-plate collectors. Chung et al. (1963) reported experiments in which a cooler was operated directly from a flat-plate collector. The collector was a two-cover flat-plate water heater and collector area was one-third to one-fifth of the required area; a thermal amplifier was used to simulate accurately a larger collector. No auxiliary energy supply (in the usual sense) was used, and no energy storage was provided. The cooler was an Arkla machine of nominal three-ton capacity that was designed for steam heating of the generator; external modifications were made to the generator to allow solar heated water to be supplied to it. The cooler capacity was limited to about two-thirds of its nominal capacity by this change in its operation, and effective collector area was selected so that the maximum (noon) output matched the reduced cooler capacity. The effective collector area was readily changed in these experiments by control of the thermal amplifier.

Figure 13.2.1 shows performance of this system for two days, one cloudy and one clear. Time dependence of cooling is clearly shown, as well as the limited duration of the operation of the cooler in this system that includes no storage. Table 13.2.1 shows the integrated results of a number of days of operation of the system.

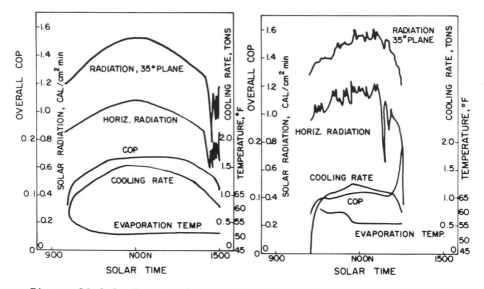

Figure 13.2.1 Two days' experimental results of operation of a flat plate collector, LiBr-H$_2$O cooler system [from Chung et al. (1963)]. On left, a clear day operation, with collector area 342 ft^2 (31.8 m^2); on right, a cloudy day operation with 377 ft^2 (35.0 m^2) collector.

From these and other experiments, it is clear that Li Br-H$_2$O absorption air conditioners can be adapted for solar operation. Without generator modification, they can be operated at reduced capacities. They can be modified to operate at nominal capacities with energy supply to generator by hot water. Part load operation can be accomplished at little loss of COP, with reduced dehumidification. Generator temperatures

TABLE 13.2.1 Summary of Results of Collector-Cooler Experiments, Integrated over a Day in Each Case [from Chung et al. (1963)].

Run No.	Duration of Run (Solar Time)	Day's Total Radiation on Horiz. Plane, Btu/ft²	Total Radiation Rec'd. by Collector, Btu/ft²	Equiv. Plate Area, Ft²	Cooling, Btu/ft²	Overall COP	Ave. Evap. Temp., °F	Av. Temp. Coolant, to Cooler, °F	Comments
50	9:30-13:58	2110	1214	527	157	0.13	50	75	Partly cloudy from 12:20 on
51	10:12-14:58	2150	1485	426	197	0.13	52	75	Scattered clouds from 12:15 to 12:50
52	10:20-14:55	2130	1397	468	160	0.12	50	75	Clear day
53	10:32-13:43	2000	1235	344	190	0.15	49	75	Partly cloudy from 11:35 on
54	10:33-14:47	1970	1424	393	184	0.13	55	82	Partly cloudy all day
55	9:48-15:44	1950	1739	389	195	0.11	55	83	Scattered clouds from 12:50 on
56	9:27-15:02	1980	1842	387	236	0.13	56	82	Clear day
57	10:12-13:25	1760	1035	379	109	0.11	57	82	Cloudy during run
58	9:18-15:00	1890	1638	342	250	0.15	54	81	Clear until 14:03

311

required for these commercially manufactured $LiBr-H_2O$ air conditioners are in a range suitable for flat-plate collectors (with the collectors operating at temperature levels above ambient about the same as those for winter heating operation). However, cooling water is required for the absorbers and condensers, and in most applications cooling towers will be needed.

13.3 ECONOMICS OF SOLAR HEATING AND COOLING

Economic considerations of combined solar heating and cooling operations are basically the same as outlined in section 12.5 of the previous chapter for heating alone, with additional factors taken into account. These additional factors are of two kinds: those concerned directly with costs, and those that affect thermal performance and thus costs.

The addition of cooling to the heating and hot water operations noted in the previous chapter has at least two effects on the performance of a system. First and most important, the collector delivers additional useful energy in the summertime, with net effect (considering the collector alone) of reducing the cost of delivered energy. Second, the use of the collector for air conditioning may reduce significantly the amount of energy it delivers for service hot water during the summer months. An analysis of the thermal performance of a system for the full range of heating and cooling seasons can show the magnitude of these and other changes on the system performance.

The cost analysis for a system should serve two purposes: to optimize the system and to compare it with conventional means of accomplishing the same ends. These comparisons should be made with systems that would most likely be used as alternatives to solar heating and cooling. At the present time, the comparison could be between a furnace and a mechanical air conditioner on the one hand, and the solar energy, auxiliary, and absorption air conditioner system on the other. As the costs of absorption coolers are significantly higher than those of mechanical coolers, an additional increment of first cost has to be charged to the solar system. [This is taken, for example, as $1000 in the Löf and Tybout (1973) study noted below, and in the example outlined in section 13.7.] On the other hand, power costs should be much lower with the solar energy systems. The comparison may then be made on the basis of total annual costs of ownership and operation for these two systems (or for other systems that may be considered).

Equations of the type of (12.5.1) and (12.5.4) are applicable to the combined operation of solar heating, hot water, and cooling. The major necessary modification to Eq. (12.5.4) is to add in the incremental cost of the absorption refrigerator, $C_{R,i}$.

$$C_{T,a} = (C_C A_C + C_{ST} + C_E + C_{R,i})I + PC_P + Q_A C_F \quad (13.3.1)$$

As in the case of heating alone, the cost of the furnace may be excluded in making cost comparisons when it is an item of equipment that is common to the systems being compared. Different amortization rates may be assigned to different items of equipment, if appropriate.

Löf and Tybout, in a paper presented in Paris (1973), have described a follow-on study to their economic study of solar heating that was discussed in the previous chapter. In this new work, they make use of a thermal model for combined operations to estimate annual performance for heating and air conditioning, and then make cost estimates for optimizing major solar system design parameters and comparing solar and conventional systems. Their system is shown schematically in Figure 13.3.1. The cooler is a three-ton absorption machine with hot water supplied to the generator at 93 °C. It operates at full capacity at a COP of 0.6 whenever it is on, if the tank temperature is above 82 °C and below 93 °C. Auxiliary energy is used to raise the temperature of water flowing to the generator to 93 °C when the temperature in the storage tank drops below 82 °C. The entire energy needs for operation of the air conditioning are supplied from auxiliary, and the storage tank is bypassed when the tank temperature is below 82 °C. Cooling and heating loads are for the same two building designs as were assumed in the earlier Löf and Tybout study of house heating (i.e., 15,000 and 25,000 Btu/degree-day heating loads, corresponding to 0.33 and 0.55 kW/°C atmospheric temperature below 18 °C). Cooling loads were based on estimation of loads caused by infiltration (sensible and latent loads), internal heat generation, conduction through walls, and solar gain through windows.

The analysis of heating and cooling was used to study four system design parameters, including collector area, storage volume, collector tilt, and number of glass covers. As with heating, meteorological data from eight locations in the United States were used. Comparisons were made of costs of delivered solar energy for solar heating and hot water, for solar cooling and hot water, and for combined solar heating, cooling and hot water. Costs of delivered solar energy were computed as functions of several design parameters, including number of covers, volume of water storage per unit collector area, and collector area.

The optimum number of glass covers is shown to be two or

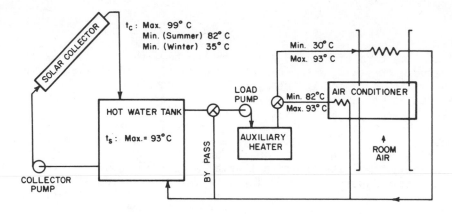

Figure 13.3.1 Schematic of the heating-cooling system modeled
by Löf and Tybout (1973).

three for the combined heating and cooling operation. The optimum amount of storage for heating and for the combined operations is about 50 kg/m². Increasing the storage capacity at
fixed collector area results in a slight increase in fraction
of heating load carried by solar, and a smaller increase in
combined loads carried by solar, for the Albuquerque climate.
The optimum collector area for this house, the Albuquerque climate and set of cost assumptions, is smallest for heating.
largest for cooling, and an intermediate value for the combined
functions. Other locations show quite different results;
Miami, for example, shows no minimum in the cost versus collector area curve for heating, and the cooling and combined systems have nearly identical cost curves.
        Löf and Tybout also compared the cost of least cost solar
energy with costs of energy from conventional sources, for heating, cooling and combined operations, in each of the eight locations. Their comparisons show, based on their cost assumptions
of $21.50/m² of collector, 20-year life, 8% interest:

        a. That in cities with high cooling loads (Albuquerque,
Miami, Charleston, Phoenix and Omaha), the cost of solar energy
for the combined operations is less expensive than for either
heating or cooling alone.
        b. That in Boston, Santa Maria and Seattle the cost of
solar energy for heating alone is less than the cost of solar
energy for the combined operation.

13.4 COMBINED SOLAR HEATING AND COOLING SYSTEMS

As has been pointed out in previous sections, we do not yet have an adequate basis of experience on which to base design of combined systems of the type under consideration here. Analytical tools that can study many designs are now being developed, and used in the planning of experiments that should soon be producing useful information. Meanwhile, several important facets of system and building design are worth noting.

One of the important considerations in combined heating and cooling systems is the relative importance of the summer and winter loads. Either one may dictate the needed capacity of the collector, and consequently its size and design. Obviously, climate is a major determining factor, and cooling requirements will dominate in climates like those of Phoenix and Miami. Commercial buildings are likely to have design fixed by cooling loads, even in more northern climates. Also important are the building design features that can affect relative energy requirements for the two loads. These features include, for example, fenestration, shading by sunshades and foliage, and building orientation. Less obvious, but of equal importance, is the performance of the heating and cooling system. For example, a poor absorption cooler will require a larger collector area than one that has a high COP, and thus could shift the determination of collector needs from winter heating to summer cooling.

The location of storage, whether inside or outside the building, will have a small effect on heating or cooling loads. If heat is to be stored, and if the storage unit is inside the structure, heat losses from storage become uncontrolled gains during the heating season, and additional loads during the cooling season.

If collectors are part of the envelope of the building, back losses from the collector will also become uncontrolled gains during heating and additional loads during cooling.

As with solar heating alone, the major design problem is the determination of optimum collector area, with underdesign leading to excessive use of auxiliary energy, and overdesign leading to lowering of the use factor on the capital-intensive solar energy system. It should also be noted that absorption air conditioners are now significantly more expensive than mechanical air conditioners. In climates where annual cooling loads are low, the use of absorption coolers will lead to higher cooling costs because of low use factors on the coolers.

It is not now possible to draw any general conclusions from these considerations. At the present time, the best procedure

for designing a system is to model it, "operate" it in the climate in question, and determine the best system design from
thermal performance and estimated costs.

13.5 MODELING OF ABSORPTION COOLERS

There are at least two possible methods available for the operation of absorption coolers with solar energy. The first is to
operate the coolers at (or near) design conditions, and supply
auxiliary energy if solar energy or energy from storage is not
adequate to operate at design rates. In that case, energy must
be available from the collector or storage at a temperature corresponding to the design input temperature of the fluid supplying heat to the generator (e.g., at 90 °C for one LiBr-H$_2$O cooler being modified in 1974 for solar operation), and energy collected at lower temperatures is useless in providing solar cooling. If the cooler is to be so operated, it may suffice to
model the cooler by assuming constant energy input, constant coefficient of performance, and constant heat transfer in the
evaporator when the machine operates. Control would be on-off,
and the heat capacity of the building and deadband of the controller should be considered in modeling these systems.

It is also possible to operate absorption coolers at less
than full capacity, by supplying them with heat transfer fluid
(e.g., water) at a lower temperature than that called for by
design conditions. Under these conditions, the coolers operate
at less than rated capacity [see Duffie and Sheridan (1965)],
and at a capacity determined primarily by the temperature of
the hot liquid to the generator and the temperature of cooling
water to the absorber and condenser.

To model this component, it is possible to use the complete
set of energy balances, material balances, rate equations, and
equilibrium relationships for each of the components in the cooler. It is also possible to derive empirical "black box" repre-
sentations of coolers. These can be based on operating experience with the machine under consideration, and on estimates of
the effects of changing operating conditions (particularly water
temperatures to the generator and absorber-condenser) on cooler
performance. An example of this approach is indicated in Figure
13.5.1, which shows the estimated capacity dependence of a modified Arkla LiBr-H$_2$O cooler on temperature of hot water to the
generator, and temperature of cooling water to the absorber and
condenser. These relationships, worked out by Butz (1973), are
based on operating experience with standard residential-type
Arkla machines, with estimates of effects of modification to
substantially reduce submergence (i.e., by addition of a

mechanical pump) and to achieve better heat transfer from hot water to the solution in the generator. Constant COP at 0.65 is assumed, based on measurements of these and other LiBr-H_2O coolers, and operation is assumed to be limited from 0.35 to 1.15 of the rated capacity. This model consists of three parts, where

$$CAP = 0.03114 \ T_{supply} - 0.04662 \ (T_{coolant} - 18.3) - 1.56 \tag{13.5.1}$$

$$0.35 < CAP < 1.15 \tag{13.5.2}$$

$$COP = 0.65 \tag{13.5.3}$$

This model illustrates the kinds of empirical relationships that can be developed and that are useful in the system analyses; it will certainly be modified on the basis of more experience. It is the model used in the example in section 13.7.

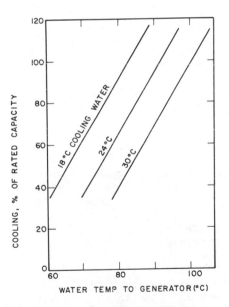

Figure 13.5.1 An example of an empirical representation of estimated dependence of cooling rate of a LiBr-H_2O cooler on temperature of the heating water to the generator and cooling water to the absorber and condenser.

Operation of LiBr-H_2O coolers at part load appears to have
significant advantages for solar operation. It can utilize some
energy collected at lower temperatures than that required for
operation at full capacity. Part-load operation may be sufficient to meet building needs during periods of low cooling requirements. When integrated over the cooling season, significantly less auxiliary energy may be required if part-load operation is allowed than if on-off operation at full load is practiced; when used with hot water energy storage systems, it permits operation of storage over a wider range of temperatures
below the boiling point, and thus reduces the need for pressurizing the collector-storage system. On the other hand, part-
load operation is generally not practiced in conventional (fuel-
fired) systems, as it may lead to inadequate dehumidification
of the building air. This consideration may dictate the need
to redesign evaporators for part-load operation, if collector
and system performance are such that part-load operation remains
desirable.
     Thus, it can be observed that commercially manufactured ab-
sorption air conditioners have been designed for operation with
conventional energy sources. The transient and temperature-
limited nature of the solar energy supply impose a new set of
design considerations that have not yet been fully considered
in cooler design. We can now identify the generator and evaporator as critical components. We can also speculate that the
sensible heat exchanger between the generator and absorber may
also be critical in obtaining higher COP; this may be important
for applications in climates and buildings where cooling loads
are more important than heating loads.

13.6 *MODELING OF HEATING AND COOLING SYSTEMS*

The general procedure for modeling these systems is the same as
that for heating alone, but additional components must be considered, including the cooling equipment. The time dependence
of cooling loads is important, and the building should be adequately modeled to reflect this dependence. In contrast to solar heating, where degree-day models may be adequate, more
detailed consideration of the building as a component in the
system is needed. Heat and moisture generation in the building,
infiltration, and solar radiation received through fenestration
may all be significant in determining sensible and latent loads.
Thus, for simulation of solar cooling, an additional meteorological parameter, the wet bulb temperature is needed (as well as
solar radiation, wind speed, and dry bulb temperature needed for
heating calculations). Reference is again made to the *ASHRAE*

Handbook of Fundamentals (1972) and various programs that are available for computation of cooling loads.

13.7 PERFORMANCE AND COST CALCULATIONS OF A SOLAR HEATING AND COOLING SYSTEM

In Chapter 12 we outlined a recent simulation study by Butz (1973) of the thermal performance of a building equipped for solar service hot water heating, space heating, and air conditioning. In that discussion, we did not include the cooling operation. Here we include the air conditioner model in the system, which is shown schematically in Figure 13.7.1. The building and other features of the system are as previously outlined, with the addition of the air conditioner, cooling tower, and modified controls for operation of the air conditioner. As before, the system was operated in the Albuquerque climate, and the main design parameter studied was collector area. The new components and controls, and the manner in which they were modeled, can be summarized as follows:

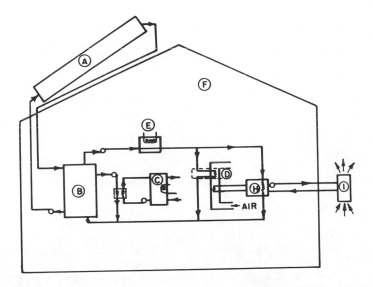

Figure 13.7.1 The Albuquerque house system modeled by Butz (1973), with air conditioning. A, Collector; B, storage tank; C, hot water system; D, space heating heat exchanger; E, main auxiliary heater; F, the building; H, absorption air conditioner; I, cooling tower.

H. The absorption air conditioner is taken as a modified
Arkla three-ton unit, with several significant changes in its
design. The generator is assumed to be operated with hot water
and to have adequate heat transfer area to operate at full capacity. Furthermore, it was assumed to operate at submergence
of about one-third of the values encountered in the "standard"
Arkla units. This operation lowers the generator temperature
several degrees and implies a variable speed mechanical pump
to circulate the absorbent in the cooler. The cooler was
modeled by the empirical relationships shown in Figure 13.5.1,
which are based on experiments with these machines (and with
other LiBr-H₂O coolers), and the predicted changes in perfor-
mances based on the modifications noted above.

I. The cooling tower was assumed to be such that the water
temperature leaving the tower and going to the air conditioner
was 6 °C above the ambient wet-bulb temperature. (This is a
conservative assumption, in the sense that at full load this approach to wet bulb is reasonable and at part load closer approaches to wet-bulb temperatures would be possible.)

The control system, not shown in Figure 13.7.1, controls
the cooling operation in a manner analogous to the control of
the heating system. That is, as the temperature rises, the
cooler operates first from the water from the storage tank,
and then from one or both stages of auxiliary input. The control functions for both heating and cooling are shown in Figure
13.7.2. The collector is controlled independently, as previously noted.

Operation of this system was simulated in the Albuquerque
climate for each of four collector areas, to provide heating,
cooling, and hot water. System design parameters were as indicated in Table 12.8.1. In each case the major energy quantities of interest were the total energy needs and the amounts
supplied by solar energy and by the auxiliary source. Figure
13.7.3 shows, as an example, the daily averages of energy quantities in the system for each month, for a collector area of
32.5 m².

This is the same collector area as that of Figure 12.8.3.
The addition of substantial energy from solar energy to operate
the air conditioner is obvious. The use of solar energy to
operate the air conditioner leads to the use of some auxiliary
energy to operate the service hot water system in the summer.
The annual pattern of energy use is now bimodal; the collector
is nearly adequate to meet energy needs in spring and fall,
but significant auxiliary energy is required in summer and winter.

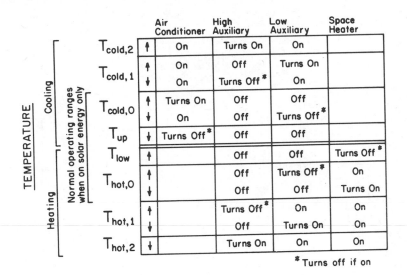

		Air Conditioner	High Auxiliary	Low Auxiliary	Space Heater
$T_{cold,2}$	↑	On	Turns On	On	
$T_{cold,1}$	↑	On	Off	Turns On	
	↓	On	Turns Off *	On	
$T_{cold,0}$	↑	Turns On	Off	Off	
	↓	On	Off	Turns Off *	
T_{up}	↓	Turns Off *	Off	Off	
T_{low}	↑		Off	Off	Turns Off *
$T_{hot,0}$	↑		Off	Turns Off *	On
	↓		Off	Off	Turns On
$T_{hot,1}$	↑		Turns Off *	On	On
	↓		Off	Turns On	On
$T_{hot,2}$	↓		Turns On	On	On

*Turns off if on

Figure 13.7.2 Schematic of the control system functions for the heating and cooling system as modeled by Butz with a two stage auxiliary source. ↑ means as room temperature rises past thet control temperature, the indicated change is made. ↓ means as temperature drops past the control temperature, the indicated change occurs.

Computed integrated energy quantities by months for two collector areas are shown in Tables 13.7.1 and 13.7.2. These tables are analogous to those of Chapter 12 for space heating and hot water.

The computed results for total cooling load were dependent to some degree, on the collector area. This dependence is attributed to two factors. First, the control system was such that, with larger collectors, conditions in the building were maintained more nearly at the desired levels. Systems with smaller collectors allowed more excursions outside the desired range, and thus required less total energy for operation of the cooler. Second, the storage tank capacity, and thus thermal losses from the storage tank, were functions of collector area; as thermal losses from the tank were to the interior of the building, they add to the cooling load to an extent determined by collector area and, thus, storage tank size. Also, the mean temperatures of the storage tank increases as the collector area

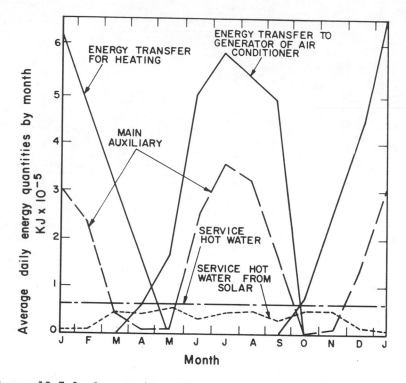

Figure 13.7.3 Computed monthly variation of average daily energy quantities for the Albuquerque House, for a collector area
of 32.5 m² [adapted from Butz (1973)].

increases. Tables 13.7.1 and 13.7.2 show the cooling loads computed for two collector areas; the cost calculations noted below are based on the highest cooling load (i.e., on the 88.3 m²
collector) and thus include highest auxiliary energy requirements.
     In this example, there was a need for both heating and
cooling only in the month of April, and the needs for both of
the functions were low. Comparison of these tables with Tables
12.8.2 and 12.8.3 shows that the annual energy gain, with the
addition of the cooling operation, increased for the 32.5 m²
system from 62.1 to 94.2 × 10⁶ kJ. For the 60.4 m² system, the
annual gain increased from 77.4 to 135.6 × 10⁶ kJ. The additional gain from the collector is significant indeed.

TABLE 13.7.1 Computed Monthly Performance of the Albuquerque Model House [from Butz (1973)].*

Month	Heating			Cooling			Hot Water			Q_{SOL}	Combined Function Total		
	Total	Solar	Aux.	Total	Solar	Aux.	Total	Solar	Aux.		Total	Solar	Aux.
Jan	18.7	9.6	9.1	0	0	0	2.0	0.3	1.7	28.6	20.7	9.9	10.8
Feb	12.7	6.6	6.1	0	0	0	1.8	0.4	1.4	21.2	14.5	7.0	7.5
March	8.6	7.7	0.9	0	0	0	2.0	1.4	0.6	32.4	10.6	9.1	1.5
April	4.5	4.5	0	1.8	1.7	0.1	1.9	1.3	0.6	25.3	8.2	7.5	0.7
May	0	0	0	4.9	4.6	0.3	2.0	0.4	1.6	24.3	6.9	5.0	1.9
June	0	0	0	14.5	7.7	6.8	1.9	1.0	0.9	24.1	16.4	8.7	7.7
July	0	0	0	17.5	6.9	10.6	2.0	1.4	0.6	23.8	19.5	8.3	11.2
Aug	0	0	0	15.9	6.4	9.5	2.0	1.6	0.4	23.4	17.9	8.0	9.9
Sept	0	0	0	14.1	9.5	4.6	1.9	1.0	0.9	27.5	16.0	10.5	5.5
Oct	2.0	2.0	0	0	0	0	2.0	1.4	0.6	23.8	4.0	3.4	0.6
Nov	6.9	6.7	0.2	0	0	0	1.9	1.4	0.5	28.3	8.8	8.1	0.7
Dec	12.1	8.3	3.8	0	0	0	2.0	0.6	1.4	23.6	14.1	8.9	5.2
	65.5	45.4	20.1	68.7†	36.8	31.9	23.4	12.2	11.2	306.3	157.6	94.4	63.2

*Cooling collector area = 32.5 m², energy quantities = kJ × 10⁻⁶.
†The total computed cooling energy requirement varied with collector area. In the cost analysis the highest values were used. (See text.) Annual η = 94.4/306.3 = 0.31.

323

TABLE 13.7.2 Computed Monthly Performance of the Albuquerque Model House [from Butz (1973)].*

Month	Heating			Cooling			Hot Water			Q_{SOL}	Combined Function Total		
	Total	Solar	Aux.	Total	Solar	Aux.	Total	Solar	Aux.		Total	Solar	Aux.
Jan	18.7	15.0	3.7	0	0	0	2.0	0.8	1.2	53.3	20.7	15.8	4.9
Feb	12.7	9.6	3.1	0	0	0	1.8	1.0	0.8	39.3	14.5	10.6	3.9
March	8.6	8.6	0	0	0	0	2.0	1.8	0.2	60.1	10.6	10.4	0.2
April	4.5	4.5	0	2.4	2.4	0	1.9	1.8	0.1	46.9	8.8	8.7	0.1
May	0	0	0	6.0	6.0	0	2.0	1.9	0.1	44.2	8.0	7.9	0.1
June	0	0	0	15.8	13.4	2.4	1.9	1.4	0.5	44.6	17.7	14.8	2.9
July	0	0	0	18.4	13.4	5.0	2.0	1.6	0.4	44.2	20.4	15.0	5.4
Aug	0	0	0	16.9	12.5	4.4	2.0	1.7	0.3	43.5	18.9	14.2	4.7
Sept	0	0	0	15.5	12.4	3.1	1.9	1.4	0.5	51.0	17.4	13.8	3.6
Oct	2.0	2.0	0	0	0	0	2.0	1.7	0.3	44.2	4.0	3.7	0.3
Nov	6.9	6.9	0	0	0	0	1.9	1.9	0	52.5	8.8	8.8	0
Dec	12.1	11.1	1.0	0	0	0	2.0	1.1	0.9	43.9	14.1	12.2	1.9
	65.5	57.7	7.8	75.0†	60.1	14.9	23.4	18.1	5.3	567.7	163.9	135.9	28.0

*Cooling collector area = 60.4 m²; energy quantities = kJ × 10⁻⁶.
†The total computed cooling energy requirement varied with collector area. In the cost analysis the highest values were used. (See text.) Annual η = 135.9/567.7 = 0.24.

324

Figure 13.7.4 shows a cumulative plot of energy quantities by months for the system with the 32.5 m² collector, and is analogous to Figure 12.8.4. For this purpose, the year starts on October 1. Figure 13.7.5 indicates the dependence of annual collector efficiency and annual energy gain as a function of collector area for the system with cooling.

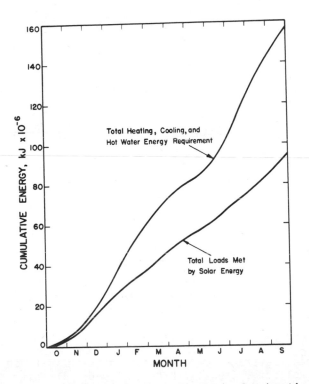

Figure 13.7.4 Cumulative total heat needs for heating, air conditioning and hot water, and energy supplied by the collector for the Albuquerque House, for a collector area of 32.5 m² [from Butz (1973)].

The comparisons against performance of systems with comparable areas of collector again show the marked increase in annual output and annual efficiency, when summer use of the collector for cooling is added. However, as before, the larger the collector, the more time it has excess capacity.

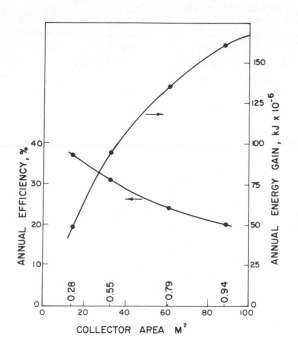

Figure 13.7.5 Annual energy gain and annual efficiency of solar
energy collection for heating, hot water and air conditioning,
as a function of collector area, for Butz's Albuquerque House.
Numbers under data points indicate fraction of the total annual
load carried by solar energy.

The monthly averages of collector efficiency are shown for
each of the collector areas in Figure 13.7.6. The smaller systems always have higher efficiencies than the larger ones, since
they operate at lower mean temperatures. The difference is
least pronounced in the summer, when the temperature of the storage tank is essentially constrained between about 70 and 110 °C.
It is most pronounced in the spring and fall, when the collector capacities most exceed the loads. While this is a specific
example, the same general trends can be expected for any system
as it will be overdesigned for parts of the year.
The thermal performance information obtained from the simulation shows how the system should perform, for this particular
year in Albuquerque. As with the system without cooling, this
information allows determination of the least-cost combination

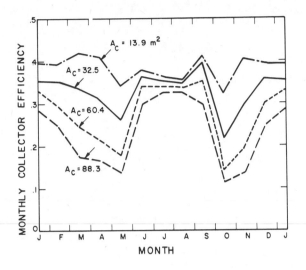

Figure 13.7.6 Monthly collector efficiencies, defined as the ratio of total collector useful gain to the total solar radiation incident on the collector for the month, for each of four collector areas [from Butz (1973)].

of solar energy and auxiliary energy, in terms of two major parameters, cost of fuel (C_F' in dollars/10^6 kJ delivered), and cost per unit area of collector, C_C.

In this example a first cost penalty of $1000 is charged to the absorption air conditioner, representing an incremental cost of the absorption cooler over the more common mechanical cooler. This is an assumption based on past costs of the types of coolers that might find application in solar air conditioning, and is subject to substantial change. Thus in this example comparisons are made between a solar and fuel operated absorption air conditioning and heating system on the one hand, and a mechanical air conditioning and fuel operated heating system on the other; other comparisons may be more appropriate for other circumstances. The cost of power for pumping has been neglected in this analysis (although the costs of conventional systems do include the cost of power at 3¢/kW-hr for operation of the mechanical air conditioner). Power costs for operation of the cooling tower were assumed small; this may not be a good assumption.

Equation (13.3.1) is the basis for the determination of costs of the combined system. Writing this in a manner analogous to Eq. (12.8.2), the total annual cost above base cost, $C_{T,a}$,

$$C_{T,a} = [(C_C + C'_{ST})A_C + C_E + C_{R,i}]I + Q_A C_F \qquad (13.7.1)$$

The same cost assumptions are made as in Chapter 12, for example,

I = 0.10185;

C'_{ST} = 8.00/m^2;

C_E = 250;

C_C = 20, 40, and \$60/m^2; and

C_F = 2, 4, and \$6/10^6 kJ delivered.

The equation used is thus

$$C_{T,a} = [(C_C + 8)A_C + 1250]0.10185 + Q_A C_F \qquad (13.7.2)$$

where C_C and C_F are the cost parameters, and Q_A is a function of A_C determined from the thermal analysis.

Figure 13.7.7 shows annual costs above base costs, with the common cost of furnace and heat distribution equipment not included, for the two collector costs and four costs of energy from fuel. The collector areas for least total cost systems are larger than those in the case of heating and hot water only, since the use factors on the collectors are higher and more useful energy is obtained throughout the year.

Comparison with the cost of conventional systems (fuel-fired furnace plus mechanical refrigeration system) can be made by assuming cost for electrical energy; if electricity costs 3¢/kW-hr, and if the overall COP of the electrical-mechanical system is 2.1, the cost of power to provide mechanical cooling for this house would be \$214 per year. At fuel energy costs of 2, 4, and \$6/10^6 kJ, the corresponding annual cost of the conventional system would be 392, 570, and \$747/yr. The total of the cost of energy from fuel and for mechanical air conditioning are shown as horizontal lines on Figure 13.7.6, although they are not functions of collector area. As with heating alone, the ranges (if any) of collector areas that will result in the combined operation being less expensive than the conventional system are evident. For example, at C_C = \$40 and C_F = \$4,

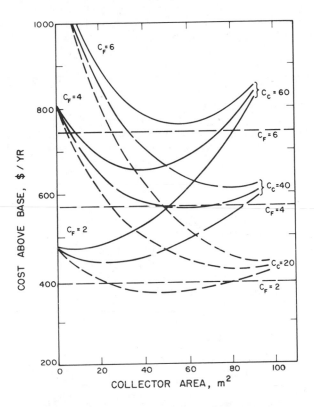

Figure 13.7.7 Total incremental annual cost, in dollars per year, for the solar heating, cooling, hot water system in the Albuquerque House. Note: At 3¢/kWh power cost, a COP of 3.0 and a motor efficiency of 0.7, the cost of operating a mechanical refrigerator to provide the same air conditioning is $214 per year; the costs of heating at three fuel costs are added to $214 and the totals for heating and cooling are shown as horizontal lines on the chart.

systems with collectors from 45 to 70 m² would be slightly less expensive than conventional systems. At C_F = $6, collector cost of slightly under $60/m², or less, would result in savings.
    In summary, this example shows how modeling and simulation techniques can provide information analogous to that obtained

from experimental systems, and how they can be used to estimate best values of design parameters. *Caution must be used*, however, in interpreting this example. It is a particular system, simulated using a particular year's meteorological data, for a particular location. Figure 13.7.6 is based on additional assumptions regarding costs and mechanical system performance. The system is one in which heating and cooling loads are reasonably well balanced. Optimum collector areas and optimum ranges for other design parameters and system configurations may be quite different in other climates and buildings. The example shows, however, the kinds of information that can be developed by simulation methods. The work of generalizing these results, checking them with experiments, and getting them in a convenient form for routine use is in its very early stages.

REFERENCES

ASHRAE Handbook of Fundamentals, New York, Am. Soc. Heating, Refrigerating and Air Conditioning Engineers, 1972.

Blytas, G. C. and Daniels, F., J. Am. Chem. Soc., *84*, 1075 (1962). "Concentrated Solutions of NaSCN in Liquid Ammonia: Solubility, Density, Vapor Pressure, Viscosity, Thermal Conductance, Heat of Solution, and Heat Capacity."

Butz, L. W., M.S. Thesis in Mechanical Engineering, Madison, University of Wisconsin (1973). "Use of Solar Energy for Residential Heating and Cooling."

Butz, L. W., Beckman, W. A., and Duffie, J. A., Solar Energy, 16, 129 (1974). "Simulation of a Solar Heating and Cooling System."

Chinnappa, J. C. V., Solar Energy, *5*, 1 (1961). "Experimental Study of the Intermittent Vapour Absorption Refrigeration Cycle Employing the Refrigerant-Absorbent System of Ammonia Water and Ammonia Lithium Nitrate.

Chinnappa, J. C. V., Solar Energy, *6*, No. 4, 143 (1962). "Performance of an Intermittent Refrigerator Operated by a Flat-Plate Collector."

Chinnappa, J. C. V., Trans. Inst. Engrs., Ceylon, 19 (1967). "The Solar Operation of a Vapour Absorption Cycle Air Conditioner at Colombo."

Chinnappa, J. C. V., Paper presented at International Solar Energy Contress, Paris (1973). "Computed Year-Round Performance of Solar-Operated Multi-State Vapour Absorption Air Conditioners at Georgetown, Guyana, and Colombo, Ceylon.

Chung, R., Duffie, J. A., and Löf, G. O. G., Mech. Engr., *85*, 31 (1963). "A Study of a Solar Air Conditioner."

Duffie, J. A. and Sheridan, N. R., Mech. and Chem. Engr. Trans., Inst. Engrs. Australia, *MC-1*, 79 (1965). "Lithium Bormide-Water Refrigerators for Solar Operation."

Duffie, J. A. and Sheridan, N. R. Supplement au Bulletin de l'Institute International du Froid, 381-387 (1965). "The Part-Load Performance of an Absorption Refrigerating Machine."

Eisenstadt, M., Flanigan, F. M., and Farber, E. A., Heating, Piping and Air Conditioning, *32*, No. 11, 120 (1960). "Tests Prove Feasibility of Solar Air Conditioning."

Farber, E. A., Paper presented at International Solar Energy Society Conference, Melbourne (1970). "Design and Performance of a Compact Solar Refrigeration System."

Lauck, F., Myers, P. S., Uyehara, O. A. and Glander, H., ASHRAE Trans., *71*, 273 (1965). "Mathematical Model of a House and Solar-Gas Absorption Cooling and Heating System."

Löf, G. O. G. and Tybout, R. A., Solar Energy, *16*, 9, (1974). "The Design and Cost of Optimized Systems for Cooling Dwellings by Solar Energy."

Sargent, S. L. and Beckman, W. A., Solar Energy, *12*, 137 (1968). "Theoretical Performance of an Ammonia-Sodium Thiocyanate Intermittent Absorption Refrigeration Cycle.

Sheridan, N. R., Paper presented at International Solar Energy Society Conference, Melbourne (1970). "Performance of the Brisbane Solar House."

Swartman, R. K. and Swaminathan, C., Paper presented at International Solar Energy Society Conference, Melbourne (1970). "Further Studies on Solar-Powered Intermittant Absorption Refrigeration."

Whitlow, E. P. and Swearingen, J. S. Paper presented at Southern Texas AIChE Meeting (1959). "An Improved Absorption-Refrigeration Cycle."

Williams, D. A., Chung, R., Löf, G. O. G., Fester, D. A., and Duffie, J. A., Refr. Engr., *66*, 33 (November 1958). "Cooling Systems Based on Solar Regeneration."

14. ADDITIONAL METHODS FOR SOLAR HEATING/COOLING

The examples of the previous two chapters illustrate the basic ideas behind solar heating and cooling systems. In addition to the examples cited, there are other experimental houses, or plans therefore, that have been developed. Some of these are briefly noted in this chapter for the purpose of illustrating the diversity of approaches that are possible to questions of solar heating and air conditioning. References are provided (some of them not readily available, unfortunately) for sources in addition to these brief comments.

14.1 COLLECTOR-STORAGE WALL SYSTEMS

The combination of collector and storage into a single structural part of the building--the south wall--was a major feature of the second house in the series of MIT houses, and is today being used in several buildings. The walls are vertical and the angle of incidence of solar radiation on them is high in the winter and low in the summer; these systems are for winter operation only.

Hollingsworth (1947) described the second MIT experimental structure, which consisted of a building oriented in an east-west direction with seven small cubicles along its length, each having its south wall made up of double glazing with a solid wall of heat storage material behind the glazing. Movable insulating shades and fans for air circulation were used to control losses and increase the rate of heat transfer from the storage wall to the room. (The designers of this experiment used separate collector and storage units in MIT House III.)

More recently, Trombe and his colleagues (1972) have built several houses at Odeillo, France, using similar principles. A cross section of one of these houses, showing the collector-storage wall, is shown in Figure 14.1.1 and a photograph of one of the houses is shown in Figure 14.1.2. The south wall is double glazed. Behind the glazing, at a distance of about 10 to 20 cm from the glass, there is a concrete wall about 20 cm thick,

332

painted black, which serves as both a radiation absorber and a heat storage medium. Openings are provided through the concrete at top and bottom, so that air circulates through the space between the glass and concrete, and through the room. This circulation is by natural convection, and no pumps or controls are used. Operating information on these houses and how it relates to the climate in the Pyrenees is not yet available.

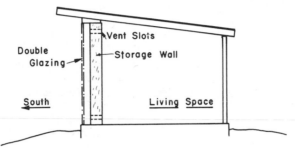

Figure 14.1.1 Schematic section of the Odeillo solar heated houses.

Figure 14.1.2 Odeillo solar heated house.

Baer (1973) has constructed buildings in New Mexico with the south walls including stacked drums of heat storage material and with movable hinged covers on the exterior to control thermal losses and radiation on the storage wall.

14.2 *COLLECTOR-RADIATOR-STORAGE SYSTEM*

Hay (1973) and Hay and Yellott (1970) describe a building designed for arid areas of the southwestern United States. The collector-radiator and storage medium are combined in the flat, horizontal roof of the one-story building. Water in a layer about 25 cm deep is contained in black plastic bags that are supported in the roof structure. Movable insulation is provided over the roof structure. It is drawn aside so that the water and plastic can absorb radiation in the daytime during the heating season, or radiate to the night sky during the cooling season. Otherwise, the insulation is located in place over the roof to prevent nighttime losses or daytime energy absorption during the heating and cooling seasons, respectively. In a test house in Arizona, movable insulation was also provided under the roof (ceiling) to control heat transfer by radiation and convection between the storage medium and the living spaces. A new house has been built in 1973 in California to study this system; data on its operation should be available in 1974.

14.3 *COLLECTOR-RADIATOR-HEAT PUMP SYSTEMS*

Yanagimachi (1958,1964) and Bliss (1964) have built and operated heating and cooling systems that use uncovered collectors as daytime collectors and nighttime radiators, "hot" and "cold" water storage tanks to supply heating or cooling to the buildings, and heat pumps to assume maintenance of adequate temperature differences between them. The Yanagimachi system was applied to a series of houses in the Tokyo area. The Bliss system was used on a laboratory in Tucson, Arizona. The systems are similar in concept, and we summarize here information on design and operation of the Bliss system.

The site of the Tucson laboratory is characterized by high radiation, low rainfall, hot summers, mild winters, and low wind velocities. These were primary considerations in design of the experimental structure, which was a one-story laboratory building having a flat roof tilted 7° to the south. The roof surface was covered with tube-in-strip copper sheeting, painted dark; it served as an uncovered solar energy collector to heat water and as a radiator to reject heat to clear night skies.

A vertical water tank, divided at its midpoint by a thermal baffle, provided storage of both hot water (in the top section)
and cold water (in the bottom section). In addition, a heat
pump was provided to transfer heat from the cold to the hot tank
sections; the evaporator coil was in the bottom of the tank and
the condenser coil was in the top.

A schematic diagram of the system is shown in Figure
14.3.1. Its operation was in any one of three modes; heating
only, cooling only, or heating and cooling.

A. In the heating-only mode, solar energy was collected
when possible, and the heated water was circulated to the bottom of the tank. Hot water stored in the top of the tank was
circulated to the radiant heating panel when the room thermostat called for heat. The heat pump operated whenever necessary
to "pump heat" from the bottom to the top of the tank, to raise
the top tank temperature to levels where building heat needs
could be satisfied.

B. In the cooling-only mode, water in the upper tank was
cooled by nocturnal radiation from the collector-radiator.
Cooling water for cooling the building was withdrawn from the
bottom of the tank, and when building needs required it, the
heat pump lowered the temperature of the water in the bottom
section.

C. In the heating and cooling mode, used in the spring and
fall, solar heated water was stored in the top of the tank, radiatively cooled water was stored in the bottom, and heating or
cooling for the building was provided from the appropriate tank
section. The heat pump operated to raise the temperature in the
top section or to lower the temperature in the bottom section
if building needs required it.

Bliss provides operating information from this building, that
shows electrical energy consumption, energy gained and rejected
by the collector-radiator, energy transferred to or from the
buidling, and other items in the energy balance. These monthly
balances are shown in Figure 14.3.2. Data from this building
show that the monthly average energy radiated from the collector
in the cooling season is, at best, about 360 Btu/ft^2/night.

Bliss points out that with systems in which cooling for the
building is provided by radiator panels, dehumidification is not
accomplished. Thus such systems, that reject heat from room air
at temperatures well above evaporator temperatures encountered
in the usual air conditioners, may be restricted to dry climates
or separate dehumidification may have to be provided.

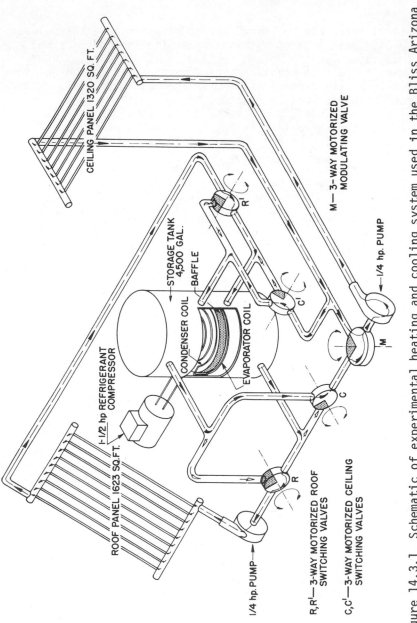

CEILING PANEL 1320 SQ. FT.

STORAGE TANK 4,500 GAL.

BAFFLE

CONDENSER COIL

EVAPORATOR COIL

1-1/2 hp REFRIGERANT COMPRESSOR

ROOF PANEL 1623 SQ.FT.

1/4 hp. PUMP

M— 3-WAY MOTORIZED MODULATING VALVE

1/4 hp. PUMP

R,R¹— 3-WAY MOTORIZED ROOF SWITCHING VALVES

C,C¹— 3-WAY MOTORIZED CEILING SWITCHING VALVES

R¹

C¹

M

C

R

Figure 14.3.1 Schematic of experimental heating and cooling system used in the Bliss Arizona laboratory building [from Bliss (1964)].

336

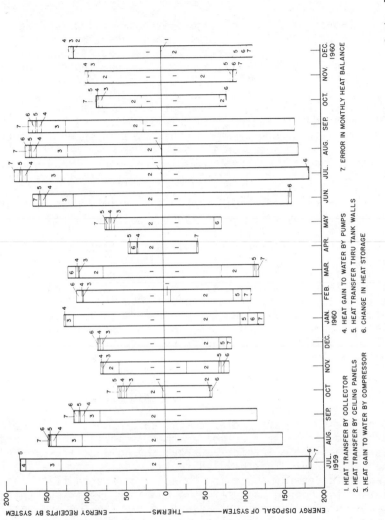

Figure 14.3.2 Monthly summary of the energy received and rejected by the water of the Arizona laboratory heating-cooling system [from Bliss (1964)]. Therm = 10⁵ Btu.

337

14.4 SOLAR ENERGY-HEAT PUMP SYSTEM

The performance of solar collectors is best at low temperatures, and the performance of heat pumps is best when the evaporator (source) temperature is highest. This has led to consideration of use of solar collectors as a source for heat pumps, for example, by Jordan and Threlkeld (1954), and by an AEIC-EEI Heat Pump Committee (1956). An office building in Albuquerque was heated and cooled by a collector-heat pump system see Haines (1956) and Bridgers, Paxton, and Haines (1957a,b)].

A schematic of one system of this type is shown in Figure 14.4.1. Storage is shown on the evaporator side of the heat pump, and a single stage heat pump is indicated. Heat storage can also be provided on the condenser side; in that case the heat pump capacity must be matched to the maximum available energy from the collector, rather than the maximum heating loads on the house. Many other variations on this basic concept are possible (the Bliss and Yanagimachi designs are developments of this basic idea). The economics of such systems has not been fully explored; while the performance and cost of the collector and heat pump are both improved over what each would be without the other, there is the necessity to pay for two relatively capital-intensive components. There may be the additional disadvantage that electrical energy would have to be used, which may lead to unfavorable load distribution at power plants; a conceivable alternative is on-site power generation (see the following section).

14.5 THERMAL AND PHOTOVOLTAIC SYSTEMS

Photovoltaic converters (solar cells) absorb most of the solar spectrum and convert a fraction of that radiation (on the order of 15% or less) to electrical energy; the balance is rejected as thermal energy. This has led to the possibility that cells be used as the energy absorbing "black" surface in flat-plate collectors that would then serve as combined electrical energy and heat source. The evident advantage lies in the combination that makes the flat-plate collector part of the protective and mounting system for the solar cells. The disadvantage lies in the fact that the performance of solar cells diminishes as their temperature increases, and permanent damage can be sustained by the cells if they are allowed to reach too high a temperature. Thus, the combined operation is limited to low temperatures (below 70 °C in the case of CdS cells).

Wolf (1972) has suggested this combination, using silicon photovoltaic cells. Böer (1973) is studying this system, using

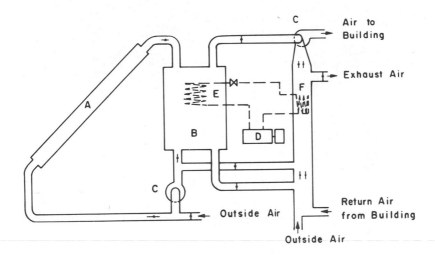

Figure 14.4.1 Schematic of a solar energy-heat pump system for heating and cooling [adapted from Jordan and Threlkeld (1954)]. Dampers are shown in position for operation of the collector and heating with the heat pump. A, collector; B, storage; C, fans; D, compressor; E, evaporator; F, condenser.

CdS cells, and is developing an experimental house at the University of Delaware that will use CdS cells in the low-temperature portions of the solar air heaters. Means are designed into the collectors to vent the spaces between the cells and covers during periods of high radiation and no heat collection (principally in summer) to protect cells from over-heating. Figure 14.5.1 is a schematic cross section of the collector-photocell assembly for the Delaware house. The house is shown in Figure 14.5.2.
    The future of such a combination depends on the successful reduction of costs of solar cells from the 1973 levels.

14.6 OPEN-CYCLE COOLING SYSTEMS

The last of these several methods for obtaining heating or cooling by solar energy is closer in concept to the absorption cooling method described in Chapter 13 than to any of the others in this chapter. Here we consider briefly several proposals, including two studied experimentally, for solar operation of open cycle absorption coolers in which water is the refrigerant.

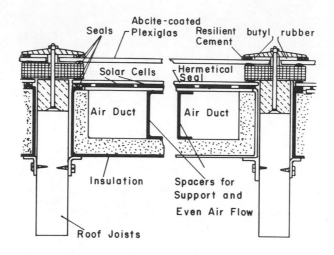

Figure 14.5.1   Cross section of the collector-CdS photovoltaic converter for the University of Delaware house [from Böer (1973)].

Figure 14.5.2   Solar One, the University of Delaware house.

   Löf (1955) suggested the system that is shown in Figure
14.6.1. The basic concept is to dehumidify room air with a
dessicant, evaporatively cool the dehumidified air, and regen-
erate the dessicant with solar energy. In this system, the
drying agent is hygroscopic triethylene glycol. The glycol is
sprayed into an absorber where it picks up moisture from the
building air. It is then pumped to a stripping column where it
is sprayed into a stream of solar heated air; at the higher
temperatures water is removed from the glycol, which is then
returned to the absorber. Heat exchangers are provided to recover sensible heat, maximize the temperature in the stripper,
and minimize the temperature in the absorber. Eliminators are
shown to remove glycol spray from the air streams. This particular solar cooling method was studied experimentally 25 years
ago, but has not been actively considered since then.

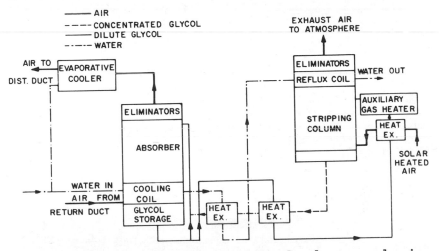

Figure 14.6.1 Schematic of triethylene glycol open cycle air
conditioning system [from Löf (1955)].

   Dunkle (1965) discusses a related method for obtaining air
conditioning in humid tropical or sub-tropical areas, which is
based on use of dessicant beds such as silica gel for air drying
(rather than the liquid absorbent suggested by Löf). The des-
sicants are regenerated by solar heated air, and a rock pile
energy storage unit is included to allow operation during times
of inadequate solar radiation. Extensive use of rotary heat exchangers is provided to maximize cycle effectiveness. The only
energy requirement, other than solar, is for movement of air.
Figure 14.6.2 shows a diagram of Dunkle's proposed method and
Figure 14.6.3 shows the cycle on the psychrometric chart, with

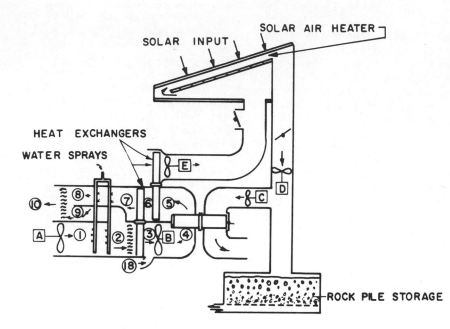

Figure 14.6.2 Schematic of open-cycle solar air conditioning system using rotary dessicant beds and heat exchangers [from Dunkle (1965)].

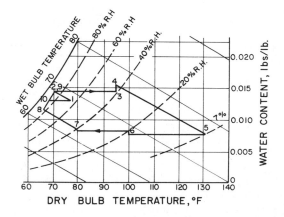

Figure 14.6.3 The cycle on a psychrometric chart [from Dunkle (1965)].

with the state number corresponding to positions on the schematic.

Neither of the above open-cycle methods are being actively pursued today. However, Baum et al. (1973) are working with a related system for solar cooling in hot, dry climates. The cycle is shown schematically in Figure 14.6.4. The absorbing liquid is a solution of lithium chloride in water. Referring to the diagram, the major steps in the process are these: starting at the absorber (1), dilute LiCl solution is transferred by pump (2) to a heat exchanger-distributor-header (3) and then to an open flat-plate collector (4), where water evaporates. The concentrated solution returns, via the heat exchanger (3) for recovery of sensible heat, to the absorber. Water is cooled in the evaporator (5), with water vapor from the evaporator going to the absorber. The chilled water from the evaporator is moved by pump (6) to an air to water heat exchanger (7) which cools the building air (in this case, without direct contact with the absorbent solution). Appropriate means are provided to deaerate the solutions, recover sensible heat, and add makeup water. The absorber is cooled with a separate cooling coil.

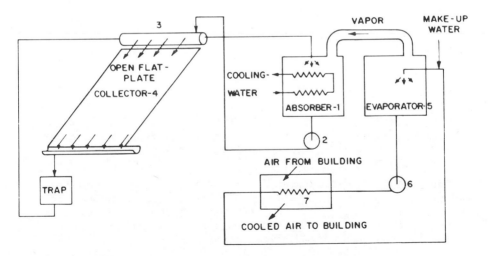

Figure 14.6.4 LiCl-H_2O open cooling system [from Baum et al. (1973)].

Close, Dunkle and Robeson (1968) reported on design and performance of a system that utilizes vee-groove selective surface air heater collectors and rock pile energy storage for solar heating of part of a laboratory building in Melbourne. During the air conditioning season the rock pile storage unit is cooled with evaporatively cooled air at night when, in that climate, the ambient wet-bulb temperatures are, at worst, 60 to 70 °F. Thus the cooling cycle was not operated with solar energy, but it used common storage equipment with the solar heating system.

REFERENCES

Collector-Storage Wall Systems

Baer, S., Report in Proceedings of the Solar Heating and Cooling for Buildings Workshop, 186, Washington (March 21 to 23, 1973), R. Allen, ed., University of Maryland, "The Drum Wall."

Hollingsworth, F. N., Heating and Ventilating (May 1947). "Solar Heat Test Structure at MIT."

Trombe, F., Personal Communication (1972).

Collector-Radiator-Storage System

Hay, H. R. and Yellott, J. I., Mech. Engr., *92*, No. 1, 19, (1970). "A Naturally Air Conditioned Building."

Hay, H. R., Mech. Engr., *95*, No. 11, (1973). "Energy Technology and Solarchitecture."

Collector-Radiator, Heat-Pump Systems

Bliss, R. W., Proceedings of the UN Conferences on New Sources of Energy, *5*, 148,(1964). "The Performance of an Experimental System Using Solar Energy for Heating, and Night Radiation for Cooling a Building,"

Yanagimachi, M., Transactions of the Conference on Use of Solar Energy, *3*, 32, University of Arizona Press, 1958. "How to Combine: Solar Energy, Nocturnal Radiational Cooling, Radiant Panel System of Heating and Cooling, and Heat Pump to Make a Complete Year-Round Air Conditioning System."

Yanagimachi, M., Proceedings of the UN Conference on New Sources of Energy, *5*, 233, (1964). "Report on Two and a Half Years' Experimental Living in Yanagimachi Solar House II."

Solar Energy-Heat Pump Systems

AEIC-EEI Heat Pump Committee, Edison Electric Inst. Bull, *77* (1956). "Possibilities of a Combination Solar-Heat Pump Unit." Based on a report by G. O. G. Löf.

Bridgers, F. H., Paxton, D. D. and Haines, R. W. Paper 57-SA-26 presented at ASME meeting (1957a). Mech. Engr., (June 1957). "Solar Heat for a Building."

Bridgers, F. H., Paxton, D. D., and Haines, R. W., Heating, Piping and Air Conditioning, *29*, 165 (1957b). "Performance of a Solar Heated Office Building."

Jordan, R. C. and Threlkeld, J. L., Heating, Piping and Air Conditioning, *28*, 122 (1954). "Design and Economics of Solar Energy Heat Pump Systems."

Thermal and Photovoltaic Systems

Böer, K., Paper presented at International Solar Energy Congress, Paris (1973). "A Combined Solar Thermal Electrical House System."

Wolf, M., Univ. of Pennsylvania, Personal Communication (1972).

Open-Cycle Cooling Systems

Baum, V. A., Kakabaev, A., Khandurdyev, A., Klychiaeva, O., and Rakhmanov, A., Paper presented at International Solar Energy Congress, Paris (1973). "Utilisation de L'énergie Solaire Dans Les Conditions Particulières Des Regions A Climat Torride Et Aride Pur La Climatisation En Eté."

Close, D. J., Dunkle, R. V., and Robeson, K. A., Mech. and Chem. Engr. Trans., Inst. Engrs., Australia, *MC4*, 45 (May 1968). "Design and Performance of a Thermal Storage Air Conditioning System."

Dunkle, R. V., Mech. and Chem. Engr. Trans., Inst. Engrs., Australia, *MC1*, 73 (May 1965). "A Method of Solar Air Conditioning."

Löf, G. O. G., Solar Energy Research, Madison, University of Wisconsin Press, 33, 1955. "House Heating and Cooling with Solar Energy."

15. NOTES ON SOLAR PONDS, SOLAR POWER, AND SOLAR DISTILLATION

The emphasis in the previous chapters has been on solar energy thermal processes, for heating and cooling applications for buildings. The component and system modeling and simulation methods outlined can be applied to other solar processes. Although we will not go into detail, these last few pages are intended to provide a starting point for further reading on several processes that have been the subject of extensive study. They include solar ponds, solar thermal power (conversion of solar to mechanical energy), and solar distillation.

15.1 SOLAR PONDS

Temperature inversions have been observed in natural lakes having high concentration gradients of dissolved salt (i.e., concentrated solution at the bottom and dilute solution at the top). This phenomenon suggests the possibility of constructing large-scale horizontal solar collectors as ponds. Solar radiation is absorbed in the lower water levels and at the bottom of the pond. The water near the bottom is at a higher temperature than the top surface, with the density of the hot concentrated lower levels higher than the density of the more dilute and cooler top levels. Experimentally, temperatures near the bottom of meter deep ponds run 70 to 80 °C and near the top 30 °C. Ponds have been studied experimentally and analytically by the National Physical Laboratory of Israel and by others. Tabor (1964) outlines the general concept and the major problems. Tabor and Matz (1965) review the NPL-Israel experimental program, and Weinberger (1964) reports a theoretical study of the physics of solar ponds.

15.2 SOLAR THERMAL POWER

Solar energy can be converted to thermal energy by flat-plate or focusing collectors. That thermal energy can then, in principle, be used to drive a heat engine, thus converting solar energy into mechanical energy. The basic problem arises from the temperature limitations of solar collectors; the solar energy supply system works best at low temperatures, while the heat engine is most efficient with energy input at high temperatures. The problem is particularly acute with flat-plate collectors, which are now limited by practical considerations to delivery of energy perhaps 100 °C above ambient temperatures; this, in turn, results in low thermal efficiencies for the heat engine. Focusing collectors, which would deliver energy at higher temperatures, are not yet practical devices. Under these circumstances, the projected costs of mechanical (or electrical) energy from solar energy have not appeared favorable.

There is a long history of experimental attempts to convert solar to mechanical energy as reviewed by Jordan (1956). Hottel (1955) made an analysis of power generation using flat-plate collectors and heat engines operating between 150 °C and 38 °C with collectors having low-reflectance and low-absorptance glass covers. Masson and Girardier (1966) reported on experimental operation of heat engines from flat-plate collectors, and Girardier and Alexandroff (1973) report on experimental installations of these systems in Africa. Currently there are several new solar power projects being sponsored by the RANN program of NSF and other agencies. For example, the Colorado State University--Westinghouse (1973) project is a general analytical study of flat-plate and focusing collector-heat engine combinations with little energy storage for generating electrical energy. The University of Minnesota--Honeywell (1973) program includes a detailed engineering study of a system based on a parabolic cylinder focusing collector.

15.3 *SOLAR DISTILLATION*

The distillation of salt water to recover potable water is accomplished by exposing thin layers of the salt water (usually in black shallow trays or basins) to solar radiation, and condensing the water vapor produced on a transparent cover in such a way that it can be collected in receiving troughs. The history of basin type stills dates back to 1872, when a still, providing drinking water for animals and used in nitrate mining in Chile, started its three decades of operation. More recently, several community scale solar stills have been built in Australia, Greece, Spain, and elsewhere, and new theoretical and practical developments have improved understanding and practice of solar distillation.

A very comprehensive review of the history, theory, applications, and economics of solar distillation has been prepared
for the Office of Saline Water by Talbert, Eibling, and Löf
(1970), and a less extensive report primarily emphasizing its
potential applications in developing economies was published
by the United Nations (1970). Dunkle (1961) analyzed the basin
type still and the multiple effect diffusion still. Cooper, in
a series of papers (e.g., 1973) has made a detailed study of
simulation of solar still performance. Proctor (1973) has experimentally investigated the possibility of augmenting solar
still output with waste heat and has found substantial increases
in yields.

REFERENCES

Solar Ponds

Tabor, H., Proceedings of the UN Conference on New Sources of
Energy, *4*, 59 (1964). "Large Area Solar Collectors (Solar Ponds)
for Power Production."

Tabor, H. and Matz, R., Solar Energy, *9*, 177 (1965). "Solar
Pond Project."

Weinberger, H., Solar Energy, *8*, 45 (1964). "The Physics of
the Solar Pond."

Solar Thermal Power

Colorado State University--Westinghouse Report to NSF, Report
NSF/RANN/SE/GI-37815/PR/73/3. "Solar Thermal Electric Power
Systems."

Girardier, J. P., Alexandroff, G. and J. Paper presented at International Solar Energy Congress, Paris (1973). "Less Moteurs
Solaries et l'Habitat pour les Zones Arides."

Hottel, H. C., Solar Energy Research, Madison, University of
Wisconsin Press, 85 (1955). "Power Generation with Solar Energy
Energy."

Jordan, R. C., Proceedings of the World Symposium on Applied
Solar Energy, Phoenix, Arizona, Stanford Research Institute,
Menlo Park, Calif. (1956). "Mechanical Energy from Solar
Energy."

Masson, H. and Girardier, J. P., Solar Energy, *10*, 165 (1966).
"Solar Motors with Flat-Plate Collectors."

University of Minnesota--Honeywell Report to NSF, Report NSF/
RANN/SE/GI-34871/PR/73/2. "Research Applied to Solar-Thermal
Power Systems."

Solar Distillation

Cooper, P.I., Solar Energy, *14*, 451 (1973). "Digital Simulation
of Experimental Solar Still Data."

Dunkle, R. V., International Developments in Heat Transfer,
Conference at Denver, Part 5, 895 (1961). "Solar Water Distil-
lation: The Roof Type Still and a Multiple Effect Diffusion
Still."

Proctor, D., Solar Energy, *14*, 433 (1973). "The Use of Waste
Heat in a Solar Still."

Talbert, S. G., Eibling, J. A., and Löf, G. O. G., Office of
Saline Water, U. S. Dept. of Interior, Research and Development
Progress Report #546 (1970), Manual on Solar Distillation of
Saline Water.

United Nations Department of Economic and Social Affairs, Publication Sales No. E.70.II.B.1, (1970), Solar Distillation.

APPENDIX A.   PROBLEMS*

The following problems, provided for those who may wish to use this as a textbook, are arranged according to the chapter in which the relevant material is found. Many can be solved quantitatively, while other are intended to be thought-provoking non quantitative questions.

Chapter 1

1.1   From the diameter and effective surface temperature of the sun, estimate the rate at which it emits energy. What fraction of this emitted energy is intercepted by the earth?

1.2   What fraction of the extraterrestrial radiation is at wavelengths below 0.5 μm? 2 μm? What fraction is included in the wavelength range 0.5 μm to 2 μm?

Chapter 2

2.1   What is the declination on January 30? March 22? August 1?

2.2   Calculate the angle of incidence of beam radiation at 1400 solar time on January 20 at latitude 35°N on surfaces with the following orientation

   a. Horizontal
   b. Tilted to south at slope of 40°
   c. At slope of 40°, but facing 25° west of south
   d. Vertical, facing south
   e. Vertical, facing west.

2.3   Determine the sunset hour angle and day length for Madison and for Miami, for the following dates: January 1, March 22, July 1.

2.4   When it is noon Pacific Standard Time in San Francisco on March 1, what is the corresponding solar time? What Central Daylight Time corresponds to solar noon in Madison on September 30?

350

*SOLUTIONS MANUAL available from the Engineering Experiment Station, University of Wisconsin, 1500 Johnson Drive, Madison, Wisconsin 53706

Chapter 3

3.1 a. For the clear day illustrated in Figure 3.2.1, deter-
         mine the hourly integrated solar radiation on a hori-
         zontal surface.

     b. For the cloudy day illustrated in Figure 3.2.2, deter-
         mine the hourly integrated solar radiation on a hori-
         zontal surface.

3.2 For Poona, India (near Bombay), estimate the monthly aver-
     age radiation for January and July from the average hours
     of sunshine data of Table 3.4.1.

3.3 Estimate the hourly direct and diffuse radiation on a hor-
     izontal surface, from the *total* radiation (i.e., the sum
     of the hourly values) determined from Figure 3.2.1 in prob-
     lem 3.1a and using the methods of section 3.5. (Compare
     the hourly results of the calculation with those that were
     summed to get the daily total.)

3.4 Estimate for May 11, for Madison, the hourly total radia-
     tion on (a), a horizontal surface, and (b), on a surface
     tilted toward the equator with a slope equal to the lati-
     tude. Weather bureau data show a total solar radiation
     for the day of 622 cal/cm^2.

     Assume: 1. Symmetry about solar noon.
              2. Diffuse radiation is concentrated in the part
                  of the sky near the sun.

     Note: For purposes of comparison, the weather bureau hour-
     ly data are:

Time	cal/cm^2	Time	cal/cm^2
5-6	0	12-1	82
6-7	7	1-2	76
7-8	22	2-3	53
8-9	39	3-4	37
9-10	54	4-5	19
10-11	77	5-6	4
11-12	82	6-7	0

3.5 Estimate the hourly direct and diffuse radiation on a hor-
     izontal surface in Madison for May 18, 1960; H = 710 cal/cm^2.
     Assume symmetry about solar noon.

3.6 a. Estimate the ratio of beam radiation on a collector
         tilted 45° toward the south to that on a horizontal
         surface, if located at Fort Collins, Colorado, on
         March 1. Make the estimation both by calculation and
         from Figure 3.6.2.

     b. Develop a chart analogous to Figure 3.6.3 for this
         situation.

3.7 For the collector and situation of problem 3.6, estimate
     the angular correction factor R based on the assumptions
     made in the development of Eq. (3.7.3), for noon on
     January 15, if the radiation is 0.7 beam with 0.3 diffuse.

     a. If there is a snow cover on the ground.

     b. If there is no snow cover.

Chapter 4

4.1 Verify the values of the blackbody spectral emissive power
     as given in Figure 4.4.1 for (a) T = 1000 K and λ = 10 µm;
     (b) T = 400 K and λ = 5 µm; and (c) T = 6000 K and λ = 1 µm.

4.2 What is the percentage of the blackbody radiation from a
     source at 300 K in the wavelength region from 8 µm to
     14 µm? (This is the so-called "window" in the earth's at-
     mosphere.)

4.3 Write a computer subroutine to calculate the fraction of
     the energy from a blackbody source at T in the wavelength
     interval a to b.

4.4 Calculate the energy transfer by radiation between two
     flat plates. The temperature and emittance of one plate
     are 500 K and 0.45 and for the other plate are 300 K and
     0.2.

4.5 Calculate three different overall heat transfer coeffi-
     cients for a plate at 50 °C when exposed to a 5 m/s wind
     and an ambient temperature of 10 °C. Base your three re-
     sults on three different approximations for the effective
     sky temperature. Plate emittance is 0.88.

4.6 Consider two large flat plates spaced L cm apart. One
     plate is at 100 °C and the other is at 50 °C. Determine
     the convective heat transfer between the plates for the
     following conditions:

     a. Horizontal, heat flow up, L = 2 cm
     b. Horizontal, heat flow up, L = 5 cm
     c. Horizontal, heat flow down, L = 2 cm
     d. Inclined at 45°, heat flow up, L = 2 cm

4.7 Compute the equilibrium temperature of a thin copper plate 1 m × 1 m × 1 mm, under the following conditions:

    a. In earth orbit, with solar radiation normal to a side of the plate. Neglect the influence of the earth. See Table 5.5.1.

    b. Just above the earth's surface, with solar radiation normal to the plate and the sun directly overhead. See Table 5.5.1.

       Assume: (a), the sky is clear and transmits 0.80 of the solar radiation; (b), the "equivalent blackbody sky temperature" is 10 °C less than the ambient temperature; (c), the ambient air temperature is 25 °C; (d), wind speed is 4.5 m/s; and (e), the earth's surface is effectively a blackbody at 15 °C.

4.8 Consider two thin circular disks, thermally isolated from each other, and suspended horizontally, side by side, in the same plane, inside a glass sphere on low conductance mounts. The sphere is filled with an inert gas, such as dry nitrogen, to prevent deterioration of the surfaces. Dimensions of the disks are identical. One disk is painted with black paint (α_b = 0.95, ε_b = 0.95) and the other with white paint (α_w = 0.35, ε_w = 0.95). The glass has a transmittance for solar radiation (τ_g) of 0.90 and an emittance for long-wavelength radiation (τ_g) of 0.88. The convection coefficient, h, between each of the disks and the glass cover is 16 W/m² °C. (Note that the disks have two sides and that the edges can be neglected.) When exposed to an unknown solar radiation on a horizontal surface, H, the temperature of the white disks (T_w) is 5 °C and the temperature of the black disk (T_b) is 15 °C. $T_{ambient}$ is 0 °C.

    a. Write the energy balances for the black and white disks assuming the glass cover is at a uniform temperature (T_c).

    b. Derive an expression for the combined convection and radiation heat transfer coefficient.

    c. Using the result of b, derive an expression giving the incident solar radiation as a function of the difference in temperature between the black and white disks.

    d. What is the incident solar radiation, H, for the conditions stated above?

Chapter 5

5.1 What are the conditions under which $\alpha = \varepsilon$?

5.2 For curve C of Figure 5.6.2b, calculate the emittance at
temperatures of (a), 300 K; (b), 500 K; and (c), 1000 K.
[Hint: Divide the spectrum into a finite number of intervals and numerically integrate Eq. (5.1.8).]

5.3 Consider a surface that has been prepared for use in outer
space and has the following spectral characteristics:

$$0 < \lambda < \lambda_c \quad \rho = 0.10$$

$$\lambda_c < \lambda < \infty \quad \rho = 0.90$$

For $\lambda_c = 1$ μm, 2 μm, and 3 μm, calculate the equilibrium
temperature of the plate. Assume the sun can be approximated by a blackbody at 6000 °K and that the solar flux on
the plate is 135 mW/cm^2. Also assume the back side of the
plate is insulated.

Chapter 6

6.1 Calculate the reflectance of one glass surface for angles
of (a) 10°, (b) 30°, (c) 50°, and (d) 70°. (Index of refraction = 1.526)

6.2 Calculate the transmission due to reflection of three glass
covers at angles of 10° and 70° and compare your results
to Figure 6.1.3.

6.3 For glass with $K = 0.20$ cm^{-1} and 0.20 cm thick, calculate
the transmission of two covers at (a) normal incidence
and (b) at 50°.

6.4 Derive an expression for the total transmission of a single
cover but do not make the assumption that $\tau = \tau_a \tau_r$ [Hint:
the method that was used to develop Eq. (6.3.1) can be used
for this problem.] Compare your solution with $\tau = \tau_a \tau_r$ for
incidence angles of 0°, 60°, and 85° for $KL = 0$ and 0.05.

Chapter 7

7.1 Verify that the overall loss coefficients for Figures
7.4.3a and 7.4.3c are correct.

7.2 Write a general computer subroutine to evaluate the overall loss coefficient for a system of n glass covers at any
angle, plate spacing, average plate temperature, ambient

temperature, and wind speed.

    a. Using Eq. (7.4.9)
    b. Without using Eq. (7.4.9)

7.3 Show that for Case b of Figure 7.4.3 the overall loss co-efficient increases from 3.98 W/m² °C to 4.05 W/m² °C when the sky temperature is reduced by 10 °C. You can do this problem with only slight modifications to the general sub-routine of Problem 7.2b.

7.4 Compare the value of U_L calculated with Eq. (7.4.9) to the graphs of Figure 7.4.4 at

    a. Wind = 5 m/s, ε_p = 0.95, T_p = 80 °C, T_a = 10 °C
    b. Wind = 10 m/s, ε_p = 0.1, T_p = 100 °C, T_a = 40 °C

7.5 For a wind of 5 m/s, an ambient air temperature of 10 °C, and an average plate temperature of 80 °C, calculate the top loss coefficient for a single plastic cover having a transmittance for infrared radiation of 0.30 and an emittance of 0.63. (These are the approximate properties of Tedlar, a duPont product). The slope is 45°, the plate spacing is 2.5 cm and the plate emittance is (a) 0.95, and (b) 0.10.

7.6 Calculate the overall loss coefficient for a flat-plate solar collector located in Madison, Wisconsin, and tilted toward the equator with a slope equal to the latitude. As-sume a single-glass cover 2.5 cm above the absorber plate (glass hemispherical emittance = 0.88), a wind speed of 4.5 m/s, an absorber plate long-wavelength emittance of 0.11, and 7 cm of rockwool insulation at the rear having a conductivity of 0.034 W/m °C. Make the calculations for absorber plate temperatures of 50, 75, and 100 °C with an ambient temperature of 25 °C. Neglect edge ef-fects and absorption of solar radiation by glass.

7.7 Consider a flat-plate collector with a fin and tube type absorber plate. Assume U_L = 8 W/m² °C, the plate is 0.5 mm thick, the tube center-to-center distance is 10 cm, and the heat transfer coefficient inside the 2 cm diameter tubes is 300 W/m² °C. Calculate the collector efficiency factor, F', for (a) copper fins, (b) aluminim fins, and (c) steel fins.

7.8 Estimate the hour-by-hour and daily collection efficiency for a flat-plate solar collector located in Boulder, Colorado, and tilted toward the equator with a slope equal to the latitude. Use the measured hourly radiation data of January 10, from Table 3.3.3.

For the collector, assume U_L = 6 W/m^2 °C, fluid to tube heat transfer coefficient is 1000 W/m^2 °C, aluminim fins and tube type construction, tube center to center distance of 15 cm, fin thickness of 0.05 cm, tube diameter of 1.5 cm, cover transmittance for solar radiation is 0.9 and independent of direction, solar absorptance of absorbing plate is 0.95, collector width is 1 m and length is 3 m, water flow rate is 75 kg/hr, water inlet temperature is constant and equal to 60 °C.

7.9 A solar water heater collector is being designed for application for space heating and cooling on an experimental building at Fort Collins, Colorado.(latitude, 41°N; longitude, 105°W). Because of architectural considerations, the slope of the collectors is to be 45° from the horizontal. One of the designs being considered calls for a three-cover collector, to be built in modules such that the absorber plate in each module measures 0.76 m wide and 3.66 m long.

The closest location for which meteorological data are available is Boulder, Colorado, a distance of about 64 km from Fort Collins. It is assumed that (in view of the proximity of location and similar location with respect to the mountain ranges) that Boulder data are transferable directly to Fort Collins. The data show that for the month of February, the average daily radiation on a horizontal surface is 348 cal/cm^2. Normal mean temperature for the month is 0.4 °C.

Collector design parameters are: F' = 0.95; G = 50 kg/m^2 hr; U_L = 2.5 W/m^2 °C, plate material is copper, coated with flat black paint; glass is low-iron window glass. Tubes are parallel to long dimension of the collector.

Calculate the expected performance of the collector for the hour from 2 to 3 p. m. (solar time) for an average February day, assuming T_a is the monthly mean temperature. For this cover system and at this time, $(\tau\alpha)$ = 0.68. Collector operation is such that the inlet water temperature is expected to be 46 °C. The following should be calculated for the hour from 2 to 3 p. m.:

    a. Useful energy gain per module of collector
    b. Outlet water temperature
    c. Efficiency

7.10 For November 2 in Madison the total solar radiation on a
horizontal surface is 8000 kJ/m² day and the air temperature is 5 °C. Estimate the steady-state efficiency and
exit temperature of a type e (Figure 7.12.1) *air heater*
at 11:30:

    a. absorber plate α = 0.95; ε = 0.2
    b. single glass cover, ε_g = 0.88; KL = .0125
    c. wind speed = 5 m/s
    d. air mass flow rate = 58 kg/hr m²
    e. air entering temperature = 38 °C
    f. plate to cover spacing = 2 cm
    g. air passage depth = 1 cm
    h. collector width = 1.3 m
    i. collector length = 3 m
    j. polyurethane foam insulation thickness = 6 cm
    k. collector tilt = 53° (south facing)

Chapter 8

8.1 It has been suggested that a solar-thermal power plant
should be based on a concentrator of the type shown in
Figure 8.2.1e, and using very large fields of reflectors
concentrating radiation onto a receiver on a central tower,
thus "transferring" solar energy from a large area into a
central location by optical means rather than piping hot
fluids. Sketch such an arrangement, showing schematically
the following dimensions: the distance L from a flat reflecting element to a spherical receiver of diameter D;
and the width of the flat reflector W. If reflector size
is limited by wind loading to W = 10 m, and if (in a large
system) L = 1 km, what (in terms of D) must be the pointing accuracy of the reflector?

8.2 Figure 8.8.1 shows an experimental flux distribution for a
cylindrical reflector, and Figure 8.8.2 the intercept factors for receivers centered at positive 0. Replot the
curve if a receiver were centered at position +0.01 m.

8.3 Consider the cylindrical focusing collector as shown in
Figure A.1. The receiver is a pipe of diameter D_p, it is
surrounded by a cylindrical glass cover of diameter D_g, and
the space between is partially evacuated. The unshaded
aperture of the reflector is W. The assembly is L ft long,
with the tubes mounted at the ends. The absorbing surface
is selectively coated.

The collector is mounted to rotate about one axis. It is used to heat and boil water. Water is pumped into one end of the pipe at 60 °C. It is heated to 285 °C and converted to steam, with dry steam, at a temperature of 285 °C, leaving the tube.

The steam from the collector is used in a "black box" heat engine-generator, which has a thermal efficiency of 20% (the ratio of electrical energy output to the useful gain of the collector).

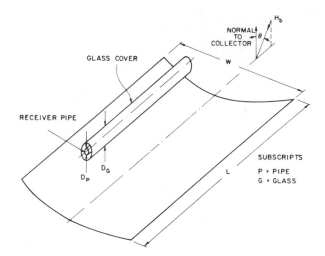

Figure A.1.

D_p = 7.5 cm
D_g = 15 cm
γ = 0.9 (intercept factor for receiver pipe)
L = overall length to be determined
W = 1.8 m
α = 0.9 (solar absorptance of receiver pipe)
ϵ_p = 0.1 (long-wavelength emittance of receiver pipe)
τ_g = 0.85 (transmittance of glass cover)
ϵ_g = 0.88 (long-wavelength emittance of glass)
ρ = 0.87 (solar reflectance of concentrator surface)
θ = 70° (angle between beam radiation and normal to collector aperture at time of interest)
H_b = 1000 W/m² (beam radiation at time of interest)

h_w = 10 W/m² °C (outside heat transfer coefficient from glass tube to air)

T_a = 25 °C (ambient air and "sky" temperature)

$\dot{m}$ = 4.4 kg/hr (mass flow rate of water into collector)

Assumptions

1. Neglect absorption of solar radiation by glass.
2. Neglect end effects.
3. Neglect temperature difference between working fluid (water) and receiver pipe.
4. Convection heat transfer between receiver pipe and glass cover has been eliminated by evacuating the glass tube. (Conduction heat transfer is not negligible since this requires a very low pressure.)
5. All radiation heat transfer terms can be linearized. (Use the *exit* fluid temperature in your linearized terms instead of some average temperature.)
6. Neglect any temperature gradients around receiver pipe and glass envelope.
7. Neglect glass thickness.

a. What is the beam radiation falling on the collector per unit length of collector?

b. Sketch the temperature distribution of the working fluid as a function of length. Label the scales and put some numbers on the temperature scale.

c. Calculate the useful gain of the collector. (Note that this is a simple calculation because the fluid temperatures have been specified.)

d. Derive an expression for the average solar energy absorbed by the receiver pipe per unit area of receiver pipe and and calculate the numerical values for the conditions given.

e. Derive expressions for the conduction and radiation heat transfer resistances per unit length from the receiver pipe to the glass.

f. Derive expressions for the conduction and radiation heat transfer resistances per unit length from the glass to the surroundings.

g. Derive an expression for the overall loss coefficient U_L in terms of the resistances of Parts d and e and the pipe diameter D (based upon the receiving pipe area).

h. Derive an expression for the useful gain of the collec-
    tor per unit length, in terms of the pipe temperature.

i. What is the length of collector necessary to heat the
    water from saturated liquid to saturated vapor?

j. What is the length of collector necessary to heat the
    liquid water from 60 °C to saturated liquid?

k. In the numerical calculation of U_L, the pipe tempera-
    ture was assumed to be uniform at 285 °C (the actual
    equilibrium glass temperature was used in the radia-
    tion resistance calculations). Since the pipe is not
    at a uniform temperature for its whole length, will
    the collector be slightly longer or shorter than cal-
    culated? Explain briefly.

Chapter 9

9.1 Develop a set of equations analogous to those in section
     9.3 for a three node (partially stratified) tank in which
     water to meet a load is replaced by water at a constant
     temperature from the mains.

9.2 Rework example 9.3.1, for two new cases:

a. The tank contains 500 kg of water, and $(UA)_s$ is
    20 kJ/hr °C.

b. The tank contains 3000 kg of water, and $(UA)_s$ is
    60 kJ/hr °C.

Chapter 11

11.1 What differences in F_R might be expected among the three
     water heater collectors shown in Figures 11.2.1 and 11.2.2?
     What practical considerations might enter into a decision
     to use one or another of these designs?

11.2 What would be the effect on the amount of solar energy de-
     livered by a collector and tank if the same water heating
     load were concentrated between the hours of 1700 and 1800,
     rather than spread out uniformly through the 24 hours of
     the day?

11.3 Based on Figure 11.5.1, what would be the relative flow
     rates through a riser one foot from the inlet and one
     eight feet from the inlet? How could these differences
     in flow be taken onto account in an analysis of a collec-
     tor, following the methods of Chapter 7?

11.4 A water heater that combines collector and storage is constructed of 150 mm diameter black tubing, placed side by side in a box insulated on its bottom so that bottom losses are small, and covered on the top by a single-glass cover. A section of the heater is shown in the sketch in Figure A.2.

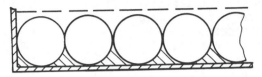

Figure A.2

The heater is filled at sunrise with water at 18 °C, and is sloped toward the equator at a tilt of 40°. Latitude is 40°N. An October day's meteorological data are as follows:

Time	Rad,kJ/m²	T_a,°C	Time	Rad,kJ/m²	T_a,°C
6-7	50	7	12-13	2710	20
7-8	160	9	13-14	2280	20
8-9	640	14	14-15	1740	21
9-10	770	16	15-16	810	22
10-11	2120	16	16-17	270	21
11-12	2770	17	17-18	20	16

Estimate the temperature history of the water in the collector through the day. State any assumptions made.

11.5 Determine the collector area to supply 75% of the hot water requirements of a residence of a family of four in Boulder, Colorado, based on the meteorological data for the week of January 8 through 14 as given in Table 3.3.3. Each person requires 45 kg of water per day at a temperature of 60 °C or above. Assume the load is uniformly distributed over the day from 7:00 to 21:00 hrs. Whenever the storage temperature is less than 60 °C, the auxiliary energy source supplies sufficient energy to heat the water, coming from the storage tank, to 60 °C.

Use the following collector and storage parameters in your analysis:

Storage mass to collector area	60 kg/m²
Collector tilt	40° to the south
Collector loss coefficient,	4.0 W/m² °C
Cover transmittance-absorptance	
product (ατ)	0.77
Collector efficiency factor,	0.95
Flow rate through the collector	50 kg/m² hr
Temperature of supply water to	
bottom of storage tank	15 °C
Tank overall loss coefficient	0.40 W/m² °C
Height to diameter ratio of cylinder	
storage tank	3
Ambient temperature around tank	21 °C

The system schematic is the same as Example 10.2.2.

11.6 Consider the water heater and storage tank of Problem 11.5. Discuss qualitatively what will happen to the system performance if the following operation or design changes are made. Consider each independently. (Note: Whenever possible, use equations to justify your *qualitative* conclusions.)

 a. The cover glass is removed (e.g., by breakage).
 b. The area of the collector is doubled.
 c. The same total load is applied for each day but it is all required between 6 p. m. and midnight.
 d. The storage tank design is changed so that water in the tank is thermally stratified rather than mixed.

Chapter 12

12.1 For the Albuquerque house modeled by Butz:

 The total annual heating load is 65.4×10^6 kJ
 The total annual hot water load is 23.1×10^6 kJ
 The total annual energy required to operate the absorption air conditioner is 82.8×10^6 kJ.

For areas of collector A_C the computer models show the following solar energy contributions to the heating load, Q_H, and the hot water load, Q_W:

A_C	Q_H	Q_W
13.9	26.2×10^6	14.1×10^6
32.5	45.5	17.1
60.4	57.6	19.8
88.3	59.1	22.0

For the following assumptions of costs, estimate the collector area leading to least annual costs:

Annual charge I = 0.10185 and 0.150
Collector cost = \$30/m^2 and \$50/m^2
Fuel costs = \$4.00/10^6 kJ

12.2 Estimate a view factor for reflected radiation from the ground for the collectors of (a) the Denver house, (b) the MIT House IV, and (c) the Odeillo house. What might the effects of snow be on the operation of these systems?

12.3 Figure 12.8.1 shows a possible solar heating system for a single family residence. The following changes have, for various reasons, been suggested:

    a. The addition of a heat exchanger between the collector and the storage tank (to permit use of antifreeze in the collector).
    b. The addition of a bypass around the storage tank to be used whenever auxiliary is on.
    c. The separation the collector into two parts, one to heat water for service hot water and the other to provide building heat, with the two systems independent of each other.

Write a brief and concise paragraph about each of these changes, stating what the major effects of the change would be. (Note: Consider each of these changes to be a separate change from the basic design shown on the diagram.)

12.4 Figure 12.8.5 shows the dependence of annual output of the collector for the Butz house as function of area. If for this house a collector area were increased from 50 to 70 m^2, an increment in collected energy would be realized. For this incremental increase in area:

    a. What is the increment in output of the solar col-
lector?

    b. What is the annual efficiency of this incremental
area?

    c. If the cost of collector is $30/m² and that is
paid off at 8% interest over 20 years, what is
the cost of the incremental energy? State what
assumptions you make in performing this calculation.

12.5 How would the solar heating system of the Odeillo house
(latitude 43°N) operate in mid July?

12.6 Figure 12.2.2 shows a basic hot water heating method with
an auxiliary supply (furnace) in series with the tank and
load. An alternative is to put auxiliary in a parallel
loop and add it to the building via a separate heat exchanger (radiator). See Figure A.3. What would be the advantages and disadvantages of such a change in design?

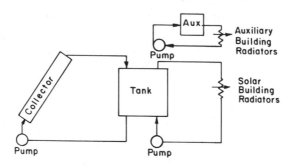

Figure A.3.

Chapter 13

13.1 If air conditioning is added to the system of Problem 12.1:

    a. For areas of collector A_c, the computer models show
the following solar energy contributions to the heating load Q_H, hot water load, Q_W, and cooling load Q_c:

A_c	Q_H	Q_W	Q_c
13.9	26.2 × 10⁶	8.6 × 10⁶	13.4 × 10⁶
32.5	45.5	12.1	36.7
60.4	57.6	18.0	60.1
88.3	59.1	21.4	81.1

For the same assumptions of costs as in Problem 12.1, estimate the collector area leading to least annual cost. Make the same assumptions of incremental costs of the air conditioner as is noted in Chapter 13.

b. Neglecting the incremental cost of the absorption air conditioner, compare the cost of delivered solar energy from the optimum area collector for heating, hot water to the cost of delivered solar energy from the optimum area collector for heating, hot water, and cooling. (Assume for this purpose that fuel costs $4.00/10^6$ kJ.)

13.2 Figure 13.5.1 shows the "black box" characteristics of an absorption air conditioner, and indicates that the capacity of the cooler diminishes as the water temperature to the generator decreases. Thus, cooling can only be obtained from solar energy if water can be delivered to the generator at approximately 80 °C (with the minimum dependent on cooling water temperature). For application at high altitudes (e.g., that of Colorado) an unpressurized storage tank can operate at a maximum temperature of 95 °C.

What advantage would there be to pressurizing the storage tank to permit its operation up to 110 °C? How can the costs and benefits of such a change be assessed?

Chapter 15

15.1 An experimental solar irrigation pump at Niamey has 27 flat-plate collectors with each unit 2 m^2. The cost of installed collector is $200 for each unit and the auxiliaries (MeCl boiler, engine, and condenser) cost $1500 to install. The average radiation on the collector at Niamey is 590 cal/cm^2/day. Estimate the cost of delivered energy, averaged over the year, per kilowatt hour, assuming:

    a. Fixed cost is 8% of the investment
       [low cost money (6%) and long life (20 yrs)]
    b. Fixed cost is 20% of the investment
       [high cost money (15%) and short life (8 yrs)].
Assume that operating cost is not a factor.

If modification of the collector costing an additional 10% and improvements in the engine and condenser costing $500 can together raise the average overall efficiency to 2.8%, what will be the costs for a and b?

15.2 Consider a solar system to supply electrical power. Determine the maximum amount of electrical power that can be produced by this system when the incident radiation is

3000 kJ/hr m^2. The characteristics of the collector are:

Collector area   10,000 m^2
Collector overall loss coefficient (U_L)   7.2 kJ/hr m^2 °C
Collector efficiency factor (F')   0.9
Mass flow rate through collector   40 kg/m^2 hr
Transmittance-absorptance product   0.75
Ambient temperature   25 °C
Fluid specific heat   4.19 kJ/kg °C

The efficiency of the heat engine is equal to one-half the
Carnot efficiency when based on the collector outlet temperature and the ambient temperature. (See Figure A.4.)

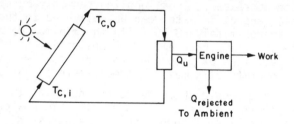

Figure A.4.

APPENDIX B. NOMENCLATURE

This Table of Nomenclature is a partial listing of symbols used in this book. Symbols that are only used infrequently are defined locally and do not appear in this list.

A area, auxiliary

C speed of light, cost

C_p specific heat

C_1 Planck's first radiation constant

C_2 Planck's second radiation constant

D, d diameter (defined locally), dust factor

E energy, equation of time

e emissive power, base of natural logarithm

F_{ij} diffuse energy leaving surface i that hits surface j without reflection/diffuse energy leaving surface i (view factor)

f focal length

$f_{0-\lambda}$ fraction of blackbody energy in region 0 to λ

F' collector efficiency factor

F'' collector flow factor

F_R collector heat removal factor

F control function

g gravitational acceleration

G mass flow rate per unit area

H total solar energy incident on the plane of measurement (usually horizontal)

H_b beam component of solar energy incident on the plane of measurement (often horizontal but may be normal)

H_d diffuse component of solar energy incident on plane of measurement (usually horizontal)

H_n	solar energy incident on surface normal to beam component
H_o	solar radiation outside the atmosphere
H_T	solar radiation on a tilted surface
h	heat transfer coefficient, Planck's constant
I	radiation intensity, annual percentage for paying interest and principle
I_{sc}	solar constant
K	length extinction coefficient
k	thermal conductivity, Boltzmann's constant
L	longitude, length, loss, load
l	length
m	mass, air mass, mean
n	index of refraction
N	number of covers
P	power
p	pressure
Q	energy per unit time
q	energy per unit time per unit area
r	radius, receiver
R	ratio of total radiation on tilted surface to that on plane of measurement, heat transfer resistance
R_b	ratio of beam radiation on tilted surface to that on plane of measurement
R_d	ratio of diffuse radiation on tilted surface to that on plane of measurement
S	absorbed solar energy per unit area
s	slope of plane from horizontal, shade factor
T	temperature
U	overall heat transfer coefficient
V	volume, velocity
v	specific volume
W	distance between tubes
z	zenith

Greek

α	absorptance, solar altitude
γ	azimuth angle, bond thickness, intercept factor
δ	declination angle, thickness (defined locally), dispersion
δ_{ij}	delta function $\delta_{ij} = 1$ when $i = j$; and $= 0$ when $i \neq j$
ε	emittance
η	efficiency (defined locally)
θ	angle (defined locally), angle between surface normal and beam radiation
λ	wavelength
μ	absolute viscosity, cosine of polar angle
ν	kinematic viscosity, frequency
ρ	reflectance, density
σ	Stefan-Boltzmann constant
τ	transmittance, time
ϕ	angle (defined locally), latitude
ω	solid angle, hour angle

Subscripts

a	air, ambient, absorbed, aperture
b	blackbody, beam, back, bond, bed
c	collector, critical, cover
d	diffuse, day
f	fin, fluid, fuel
g	glass
i	incident, inlet
l	loss
n	normal
o	overall, out
p	plate
r	radiation
s	storage, sunset, specular, scattered

T	tilt
t	top
u	useful
w	wind
z	zenith
λ	wavelength

APPENDIX C. THE INTERNATIONAL
SYSTEM OF UNITS [SI]†

SI UNITS	SOME CONVERSIONS OF UNITS

Basic units (name, symbol, quantity)

Exact conversion factors are terminated by an asterisk*

meter m--length
kilogram kg--mass
second s--time
kelvin K--thermodynamic temperature

Length m, m/s

1 ft = 0.304 8* m
1 in = 25.4* mm
1 mile = 1.609 344* km
1 ft/min = 0.005 08* m/s
1 mile/h = 0.447 04* m/s
1 km/h = 0.277 778 m/s

Derived units

All other units are derived from basic and supplementary units. Some derived units have special names.

Area m^2

1 ft^2 = 0.092 903 04* m^2
1 in^2 = 0.000 645 16* m^2
1 mile^2 = 2.589 99 km^2

Decimal multiples of units

The following prefixes are recommended for use with SI units:

Volume m^3, m^3/kg, m^3/s

(Note: 1 liter = 10^{-3}m^3)

tera T 10^{12}, giga G 10^9
mega M 10^6, kilo k 10^3
milli m 10^{-3}, micro μ 10^{-6}
nano n 10^{-9}, pico p 10^{-12}
femto f 10^{-15}, atto a 10^{-18}

1 ft^3 = 28.316 8 liters
1 U.K. gal = 4.546 09 liters
1 U.S. gal = 3.785 44 liters
1 ft^3/lb = 0.062 428 m^3/kg
1 cfm = 0.471 947 liter/s

The use of the following prefixes should be limited:

1 U.K. gpm = 0.075 768 2 liter/s
1 U.S. gpm = 0.063 090 7 liter/s

hecto h 10^2, deca da 10
deci d 10^{-1}, centi c 10^{-2}

1 cfm/ft^2 = 5.080 00 liter/s m^2
(air conditioning)

$\dagger$This table of conversion factors, constants, and properties is an edited version of a table supplied through the courtesy of the Division of Mechanical Engineering of CSIRO, Australia.

Force newton $N \equiv kg\, m/s^2$, N/m,   *Mass* kg, kg/m^3, kg/s, $kg/s\, m^2$
pascal $Pa \equiv N/m^2$

1 lbf = 4.448 22 N	1 lb = 0.453 592 37* kg
k kgf = kp = 9.806 65* N	1 oz = 28.349 5 g
1 lbf/ft = 14.593 9 N/m	1 lb/ft^3 = 16.018 5 kg/m^3
1 dyne/cm = 1 (mN)/m	1 g/cm^3 = 10^3 kg/m^3
(milli N per m)	1 lb/h = 0.000 125 99 kg/s
1 bar = 10^5 Pa	1 lb/h ft^2 = 0.001 356 2 kg/s m^2

1 lbf = 4.448 22 N
k kgf = kp = 9.806 65* N
1 lbf/ft = 14.593 9 N/m
1 dyne/cm = 1 (mN)/m
              (milli N per m)
1 bar = 10^5 Pa
1 psi = 6.894 76 kPA
1 mm H$_2$O = 9.806 65* Pa
1 in H$_2$O = 249.089 Pa
1 mm Hg = 133.322 Pa
1 at = kgf/cm^2 = 98.066 5* kPa
1 atm = 101.325* kPa

1 lb = 0.453 592 37* kg
1 oz = 28.349 5 g
1 lb/ft^3 = 16.018 5 kg/m^3
1 g/cm^3 = 10^3 kg/m^3
1 lb/h = 0.000 125 99 kg/s
1 lb/h ft^2 = 0.001 356 2 kg/s m^2

Plane angle rad
2π rad = 360* degrees

Diffusivity m^2/s

1 cST (centistoke) = 10^{-6} m^2/s
1 ft^2/h = 25.806 4 $\times$ 10^{-6} m^2/s

Energy joule $J \equiv Nm = Ws$
J/kg, J/kg°C (Note: °C = degree
                          Celsius)

SOME PROPERTIES IN SI UNITS

1 kWh = 3.6* MJ
1 Btu = 1.055 06 kJ
1 Therm = 105.506 MJ
1 kcal = 4.186 8* kJ
1 Btu/lb = 2.326* kJ/kg
1 Btu/lb°F = 4.186 8* kJ/kg°C

Density (kg/m^3)

Copper	8 795
Steel	7 850
Aluminium	2 675
Glass, standard	2 515
Concrete,typical,building	2 400
Water, 4°C	1 000
Ice, -1°C	918
Gypsum plaster, dry, 23°C	881
Oak, 14% wet	770
Pine, 15% wet	570
Pine fiberboard, 24°C	256
Asbestos cement, sheet,30°C	150
Cork board, dry, 18°C	144
Ebonite, expanded, 10°C	64
Mineral wool, batts, -2°C	32
Polyurethane, foam, rigid	24
Polystyrene, expanded, 10°C	16
Air, p$_o$, 20°C	1.204

Power watt $\equiv J/s = N\, m/s$
W/m^2, W/m^2°C, W/m°C

1 Btu/h = 0.293 071 W
1 kcal/h = 1.163* W
1 hp = 0.745 700 kW
1 Ton refr. = 3.516 85 kW
1 W/ft^2 = 10.763 W/m^2
1 Btu/h ft^2 °F = 5.678 26 W/m^2°C
1 Btu/h ft°F = 1.730 73 W/m°C
1 Btu/h ft^2 (°F/in) = 0.144 288
                          W/m°C

Viscosity Pa s = N s/m^2= kg/m s   *Thermal conductivity* (W/m°C)

1 cP (centipoise) = 10^{-3} Pa s
1 lbf h/ft^2 = 0.172 369 MPa s

Copper	385
Aluminium	211

Steel	47.6	
Ice, -1°C	2.26	

Momentum diffusivity = kinematic viscosity (m^2/s)

Concrete, typical, building	1.73	Air, 20°C, p_0	14.95×10^{-6}
Glass, standard	1.05	Water, 20°C, p_0	1.01×10^{-6}
Water, 20°C	0.596		

Asbestos cement, sheet, 30°C	0.319

Heat (thermal) diffusivity (m^2/s)

Gypsum plaster, dry, 23°C	0.170	Air, 20°C, p_0	21.2×10^{-6}
Oak, 14% wet	0.160	Water, 20°C, p_0	0.142×10^{-6}
Pine, 15% wet	0.138		
Pine fiberboard, 24°C	0.051 9	*Mass diffusivity* (m^2/s)	
Cork board, dry, 18°C	0.041 8		

Mineral wool, batts, -2°C	0.034 6

Water vapor in air at 20°C, p_0 26.1×10^{-6}

Polystyrene, expanded, 10°C	0.034 6

Surface tension (N/m)

Ebonite, expanded, 10°C	0.030 3		
Air, p_0, 20°C	0.026	Water/air, 20°C, p_0	0.0728

Polyurethane, foam, rigid	0.024 5

Ultimate tensile strength

Specific heat (kJ/kg°C)   Mild steel: about 450 MPa

Water, 20°C, p_0	4.19
Ice, -21°C to -1°C	2.10
Steam (c_p), 100°C, p_0	1.95
Air (c_p), 20°C, p_0	1.012
Concrete, 18°C	0.837

MISCELLANEOUS INFORMATION

NTP (Normal temperature T_0 and pressure p_0):
$T_0 = 273.15$ K = 0°C
$p_0 = 101.325$ kPa

Heat of vaporization (kJ/kg)

Water, 20°C	2454.0
Water, 100°C	2257.0
R12, 0°C, sat.	151.5
R22, 0°C, sat.	205.4
R11, 0°C, sat.	188.9
R500, 0°C, sat.	183.0
R717, 0°C, sat.	1263.3

Standard gravity g_0:
$g_0 = 9.806\ 65$ m/s

Velocity of sound in air:
344 m/s at p_0, 20°C, 50% R.H.

Gas constants:
$R_u = 8\ 314.4$ J/kmol K universal
$R_a = 287.045$ J/kg K air
$R_v = 461.52$ J/kg K water vapor

Viscosity (Pa s)

Water, 20°C, p_0	1010.0×10^{-6}
Air, 20°C, p_0	18.1×10^{-6}

Stefan-Boltzmann constant:
$\sigma = 5.669\ 7 \times 10^{-8}$ W/m^2 K^4

Temperature K, °C (degree Celsius)

If the same temperature is a (°C)   Temperature difference:
b (K), and c (°F) then: 1°C = 1.8°F
  a °C = bK - 273.15   1°F = 0.555 556°C
  a °C = (b°F - 32)/1.8   1 K = 1.8°F
  b °F = 1.8 × a°C + 32   1°F = 0.555 556°K

SUBJECT INDEX

AUTHOR INDEX

384